T0295149

Fire-Resistant Design of Structures

This book addresses the detailed analysis and design of structures under fire loads through the basic concepts. While fire and explosion characteristics of materials are discussed in detail, an estimate of fire load and integration to fire-resistant design is the main focus. The detailed design procedures include practical examples of various design codes from around the world. The scope of *Fire-Resistant Design of Structures* includes discussions related to the estimate of fire loads, analysis and design of structural members under fire, fire protection and firefighting systems, working principles, and suitability for various industrial applications. It provides comprehensive coverage regarding the analysis and design of structural systems under fire loads, in particular, and under elevated temperatures, in general.

Features:

- Provides an understanding of fire loads, analysis, and design of various structural members
- Includes detailed design methods and model studies
- Covers in detail different types of firefighting equipment and their functions and applications

Fire-Resistant Design of Structures

Srinivasan Chandrasekaran and Gaurav Srivastava

CRC Press
Taylor & Francis Group
Boca Raton London New York

CRC Press is an imprint of the
Taylor & Francis Group, an **informa** business

MATLAB® is a trademark of The MathWorks, Inc. and is used with permission. The MathWorks does not warrant the accuracy of the text or exercises in this book. This book's use or discussion of MATLAB® software or related products does not constitute endorsement or sponsorship by The MathWorks of a particular pedagogical approach or particular use of the MATLAB® software.

Cover Image Credit: Cover image supplied by Gaurav Srivastava

First edition published 2023
by CRC Press
6000 Broken Sound Parkway NW, Suite 300, Boca Raton, FL 33487-2742

and by CRC Press
4 Park Square, Milton Park, Abingdon, Oxon, OX14 4RN

CRC Press is an imprint of Taylor & Francis Group, LLC

© 2023 Taylor & Francis Group, LLC

Library of Congress Cataloging-in-Publication Data
Names: Chandrasekaran, Srinivasan, author. | Srivastava, Gaurav, author.
Title: Fire-resistant design of structures / Srinivasan Chandrasekaran and Gaurav Srivastava.
Description: First edition. | Boca Raton, FL : CRC Press, 2023. | Includes bibliographical references and index.
Identifiers: LCCN 2022036365 (print) | LCCN 2022036366 (ebook) | ISBN 9781032358116 (hbk) | ISBN 9781032358123 (pbk) | ISBN 9781003328711 (ebk)
Subjects: LCSH: Building, Fireproof.
Classification: LCC TH1065 .C44 2023 (print) | LCC TH1065 (ebook) | DDC 693.8/2–dc23/eng/20221012
LC record available at https://lccn.loc.gov/2022036365
LC ebook record available at https://lccn.loc.gov/2022036366

ISBN: 978-1-032-35811-6 (hbk)
ISBN: 978-1-032-35812-3 (pbk)
ISBN: 978-1-003-32871-1 (ebk)

DOI: 10.1201/9781003328711

Typeset in Times
by codeMantra

Contents

Preface

Fire Resistance Design of Structures covers the necessary domain of understanding the structural design for fire. While the focus is mainly with respect to steel structures, design examples of concrete members for fire resistance design are also explained. The main strength of the book is the MATLAB codes, which can be used to estimate the material characteristics under high temperatures. MATLAB codes are useful to establish the necessary constituent properties of materials in the numerical model for advanced research under fire loads. This book explains various examples in a simple and illustrative manner, making it very convenient for self-learning even to an undergraduate student. The presented material is useful for both academic and research purposes, while consulting engineers will enjoy the design examples and explanations. Design is carried out with respect to Indian and Eurocode, helping to develop the necessary competency in subject domain. Four chapters, namely Fire and Explosion, Material Properties, Fire Design, and Fire Protection, are carefully grouped to ensure a continuous and consistent learning of the subject. The authors are well-experienced academicians and practicing consultants, whose teaching skills and consulting methods paved a classroom teaching material for the readers. Case studies discussed in the last chapter are useful interpretations and help engineers to gain important tips for fire protection.

The first chapter describes fire and explosion processes in detail, including empirical relationships that help implement various design practices that effectively address the consequences of computing such loads. Numeric examples discussed in this chapter help the learners understand the effects caused by fire and explosion. The flammability diagram presented in this chapter also helps to understand the avoidance of fire based on managing the input that results in fire. This chapter assists in comprehensively learning the subject in a simple, compact form. Material properties are crucial to investigating fire-resistant capacity and enabling an appropriate design technique. Although many modern structural systems are designed to develop their capacity from geometric form, material strength is inherent to achieving the desired capacity. In this chapter, the mechanical properties of materials, namely steel, concrete, and functionally graded materials, are discussed in detail. Material properties at elevated temperatures and various parameters that influence the strength at high temperatures are presented to enhance the understanding of the material behavior at high temperatures.

The third chapter discusses the key ideas behind the structural design for fire including presenting a view of fire as a hazard, characterization of fire as a load, temperature calculations, and structural design calculations. Some statistics related to damages caused by fire are also included to emphasize the need of considering fire as a load, although it is an accidental occurrence. It is to be noted that the probability of occurrence of fire in a building is significantly greater than that of the building experiencing a major earthquake during its design life. Hence, appropriate treatment of fire scenarios in the design of structures should be mandatory. This chapter reviews the basic concepts of combustion of fuel in building fires and provides a brief idea about calorimetry used to determine the burning behavior of the fuel materials.

It also discusses the concepts related to the quantification of the fire load. Fire development process within the compartment and various empirical fire development models that consider fire load and ventilation conditions are also discussed in detail. Additionally, the chapter also discusses temperature progression at different stages of fire in terms of standard and parametric fire time-temperature curves.

Fire protection is based on a set of well-laid principles that already exist. It is vital that they should be followed without any deviations. A set of case studies are discussed in the fourth chapter, which is important to understand the fire protection methods and practices. This chapter is coauthored by Mr. Abhay Purandare, who is a freelance, fire, and life-safety consultant practicing in Ahmedabad. Valuable contributions to the contents and case study analyses are based on his professional experience, which is sincerely acknowledged by the authors.

The immense experience of the authors, both in academia and in structural consultancies, is combined to describe the topics presented in this book. The authors express their immense gratitude to all research scholars, graduate students, and colleagues for their support in various capacities. The lead author thanks the Chairman, Centre of Continuing Education, Indian Institute of Technology Madras, for extending the administrative support in preparing the manuscript of this book. The authors also thank the administrative support extended by the Indian Institute of Technology, Gandhinagar.

Srinivasan Chandrasekaran
Gaurav Srivastava

MATLAB® is a registered trademark of The MathWorks, Inc. For product information, please contact:

The MathWorks, Inc.
3 Apple Hill Drive
Natick, MA 01760-2098 USA
Tel: 508-647-7000
Fax: 508-647-7001
E-mail: info@mathworks.com
Web: www.mathworks.com

Authors

Srinivasan Chandrasekaran is currently a Professor in the Department of Ocean Engineering, Indian Institute of Technology Madras. He commands a rich experience in teaching, research, and industrial consultancy of about 29 years. He has supervised many sponsored research projects and offshore consultancy assignments, both in India and abroad. His active research areas include dynamic analysis and design of offshore structures, structural health monitoring of ocean structures, risk and reliability, fire-resistant design of structures, use of Functionally Graded Materials (FGM) in marine risers, and Health, Safety, and Environmental (HSE) management in process industries. He was a visiting fellow under the Ministry of Italian University Research (MIUR) invitation to the University of Naples Federico II for 2 years. During his stay in Italy, he researched the advanced nonlinear analysis of buildings under earthquake loads and other impact loads with experimental validation on full-scaled models. He has authored about 170 research papers in peer-reviewed international journals and refereed conferences organized by professional societies worldwide. He has authored 22 textbooks, which various publishers of international repute publish. He is an active member of several professional bodies and societies, both in India and abroad. He has also conducted about 20 distance-education programs on various engineering subjects for the National Program on Technology-Enhanced Learning (NPTEL), Govt. of India. He is a vibrant speaker and delivered many keynote addresses at international conferences, workshops, and seminars organized in India and abroad.

Prof. Gaurav Srivastava is the Dr Vilas Mujumdar Chair Associate Professor in Civil Engineering at the Indian Institute of Technology Gandhinagar. His main research interests lie in the area of structural fire engineering. He has been involved in experimental and simulation studies related to different kinds of fires including building facade fires and is working toward the development of efficient analysis of structures under fire to enable more comprehensive performance-based design procedures and assessment of building facade systems in fire conditions. He is also interested in developing analytical and algorithmic tools for solving structural dynamics and uncertainty quantification problems.

Contributor

Abhay Purandare
Fire Safety Consultant/Advisor
Ahmedabad, India

1 Fire and Explosion

SUMMARY

Fire and explosion are important processes for the accident loads, damages caused, and complexities involved in the design to resist them. Irrespective of the type of structure, all suffer potential threats from these loads. Understanding the process helps the designers implement various design practices that effectively address the consequences of fire and explosion. This chapter describes fire and explosion processes in detail, including empirical relationships that help implement various design practices that effectively address the consequences of computing such loads. Numeric examples discussed in this chapter help the learners understand the effects caused by fire and explosion. The flammability diagram presented in this chapter also helps to understand the avoidance of fire based on managing the input that results in fire. This chapter assists in comprehensively learning the subject in a simple, compact form.

1.1 CATEGORY AND CLASSIFICATION

Fire is a rapid, exothermal oxidation of an ignition fuel, resulting in any one of the states, namely solid, liquid, or gas. Fire is triggered when a leak (or spill) of any flammable mixture occurs in the presence of a potential ignition source. One of the severe consequences of fire is releasing as an exothermal reaction. Released energy reaches its peak intensity with time. Fire can also result from an explosion caused by pressure or shock waves (Chandrasekaran, 2019). These waves propagate very fast, causing adiabatic expansion. Fire can be categorized as follows:

- **No-risk fire** is a condition where the engineering module has no energy sources that can cause a fire.
- **Low-risk fire** condition covers those building modules that store non-flammable equipment.
- **Medium-risk fire** includes those building modules that house electric power units and major testing equipment. In the context of offshore platforms, it includes non-accommodation modules.
- **High-risk fire** includes building modules storing flammable liquids. It also refers to the segments of industrial units that house high-power electric machinery and accommodation modules.

Fire can be classified based on the type of fuel burning, and it is also sometimes referred to as fuel classification. Class A includes wood, paper, cloth, trash, and plastics but not metals. Class B includes flammable liquid, gases, oil, and grease. Class C includes fully charged or connected to a live power line, and energized electrical equipment. Class D includes metals, potassium, sodium, aluminum, and magnesium.

DOI: 10.1201/9781003328711-1

Fire classification helps choose an appropriate extinguisher type. The wrong choice shall worsen the fire scenario. Fire is mainly classified as pool fire, jet fire, fireball, and flash fire, while the sub-classifications include flare, fire-on-liquid surface, and running liquid fire (Chandrasekaran, 2016). However, these sub-classifications can be treated similarly to the main modeling and numerical simulation (Chandrasekaran and Kiran, 2014a, 2014b). For example, flare is idealized as a jet fire and fire-on-liquid surface as a pool fire (Chandrasekaran and Kiran, 2015).

1.2 POOL FIRE

Pool fire occurs due to the release of liquid as a surface spill. It occurs when flammable liquid leaks from a vessel or pipeline to form a fluid reservoir, igniting. The spread of fuel or liquid over the surface causes a pool and burns above this pool with very little momentum. Thus, a pool fire can be defined as a turbulent diffusion fire burning above a horizontal pool of vaporizing hydrocarbon fuel with a very low initial momentum. It diffuses with time as the burning is due to hydrocarbon vaporization, in particular. Such fire, which happens in the open is well, ventilated (fuel-controlled) unlike fire that outbreaks within enclosures and is termed under-ventilated (ventilation-controlled). Foam is used to extinguish pool fire by creating a barrier between the gaseous fumes and the air. It prevents the mixing of fuel and oxygen and thus helps in extinguishing the fire. In offshore platforms and large liquid storage yards, pool fire is quite common. Fuel may spill accidentally due to overfilling the storage tanks or oil containers, causing a pool around. Small dikes (also called bunds) are constructed around the storage tanks to contain the spread of the pool. Firefighting jets, provided along these dikes, shall help control the spared of fire (Chandrasekaran and Kiran, 2014b). Thus, pool fire, if contained within dikes, can be controlled to avoid its further spread. The leak of fuel, resulting in pool formation, may also arise due to pipeline rupture and corrosive connection on the fuel tanks or storage containers. While the diameter of the dike and the pool are the same, the height of the dike is about one-fourth of its diameter. Usually, dikes are constructed all around the fuel tanks, leaving a sufficient circumferential offset of about one-third of its height. The following relationship holds good to compute the length of pool fire:

$$\text{or chemicals } L_f = 42 \, D_p \left[\frac{\text{Heat release rate}}{\rho_{air} \, \sqrt{9.81 \, D_p}} \right]^{0.61} \tag{1.1}$$

where D_p is the pool's diameter (m), and ρ_{air} is the mass density of air, which is 1.225 kg/m^3 under standard conditions. The burning rate is the rate of mass loss of the condensed-phase fuel. It is the evaporation rate of liquid fuel or pyrolysis rate of the solid fuel, respectively. The following empirical relationship gives the heat release relationship:

$$\dot{Q} = \dot{m} \, \Delta H_c \left[1 - e^{-k\beta \, D_p} \right] A_{dike} \tag{1.2}$$

where $\dot{Q}$ is the heat release rate (kW), $\dot{m}$ is the mass burning rate of fuel (kg/m^2-s), ΔH_c is the effective heat of combustion of fuel (kJ/kg), and $k\beta$ is the empirical constant derived based on experimental studies (m^{-1}).

Flame length, an indicator of fire intensity, is the distance measured from the average flame tip to the middle of the flaming zone at the base of the fire. It may be measured with a slant height if the flame is tilted due to the wind effect. Flame height is the average height of the flame, measured vertically. It may be lesser than the flame length if the flame is inclined under wind action. The average flame height is given by the following relationship (Budnick et al., 1997; Delichatsios, 1984):

$$H_f = 0.235 \left(\dot{Q} \right)^{\frac{2}{5}} - 1.02 \ D_p \qquad (1.3)$$

where $\dot{Q}$ is the heat release rate of the fuel.

In case of fire abutting against a wall, the wall-side heat flux is governed by the flame radiation and that in the far field by convection (Delichatsios, 1984). Hence, flame height can be used to represent the distribution of wall heat transfer, which shall depend on the gross HRR (heat release rate). Flame height and length refer to the lengths of flame in the vertical and horizontal directions, respectively. It is assumed that the flame abutting against a wall is axisymmetric and can entrain air from half of its perimeter, making the mass flow rate half of this flame. The flame height of a wall fire can be computed using the following empirical relationship (Budnick et al., 1997):

$$H_{f, \ wall \ fire} = 0.034 \left(\dot{Q} \right)^{\frac{2}{3}} \qquad (1.4)$$

where $\dot{Q}$ is the HRR per unit length (kW/m), and the factor (0.034) is entrainment coefficient. In a few cases, fire abutting the wall may grow like a line fire instead of a wall fire. The height of such flame is given by:

$$H_{f, \ wall \ line} = \frac{0.034}{2} \left(\dot{Q} \right)^{\frac{2}{3}} \qquad (1.5)$$

Fire (flame) originating from a corner of a compartment is referred to as a corner fire and is more severe than a wall fire. A corner fire may be modeled using a pool fire and specifying the center coordinates as the corner's apex (Hasemi and Tokunaga, 1983, 1984). At the start of the fire, a diffusion flame comes in contact with the wall and spreads along with the interaction of the wall and the ceiling to reach the other corner. Flames spread downwards in a non-combustible mode; or else, it grows as a rapid-fire. The height of the corner is given by the following empirical relationship (Hasemi and Tokunaga, 1983):

$$H_{f, \ corner \ fire} = 0.075 \left(\dot{Q} \right)^{\frac{3}{3}} \qquad (1.6)$$

The mass burning rate of hydrocarbon fuel can be seen in Table 1.1 (Morgan et al., 2015).

TABLE 1.1
Hydrocarbon Fuel Calorific Properties

Fuel	Mass Burning Rate (kg/m²s)	Heat of Combustion (kJ/kg)	Empirical Constant kβ
Acetone	0.041	25,800	1.9
Benzene	0.085	40,100	2.7
Benzene	0.048	44,700	3.6
Butane	0.078	45,700	2.7
Crude oil	0.034	42,600	2.8
Diesel	0.045	44,400	2.1
Diethyl ether	0.085	34,200	0.7
Dioxane	0.018	26,200	5.4
Ethanol	0.015	26,800	100
Fuel oil, heavy	0.035	39,700	1.7
Gasoline	0.055	43,700	2.1
Heptane	0.101	44,600	1.1
Hexane	0.074	44,700	1.9
JP-4	0.051	43,500	3.6
JP-5	0.054	43,000	1.6
Kerosene	0.039	43,200	3.5
Lube oil	0.039	46,000	0.7
Methanol	0.017	20,000	100
Silicon transformer Fluid (561)	0.005	28,100	100
Transformer oil, Hydrocarbon	0.039	46,000	0.7
Xylene	0.09	40,800	1.4

Example 1.1

A pool fire is inundated due to fuel leakage from a tank at a heat release rate of 2,000 kW. It may be noted that the standard measuring system of heat release rate is Watts, which is 1 Joule per s. Let the diameter of the pool formed be 2.0 m. Estimate the average height of the flame and length of the pool fire inundated.

Solution

The average flame height of pool fire is computed using Eq. (1.3).

$$H_f = 0.235 \, (2{,}000)^{\frac{2}{5}} - 1.02 \, (2.0) = 2.875 \text{ m}$$

Length of the pool fire is computed using Eq. (1.1).

$$L_f = 42 \, (2.0) \left[\frac{2{,}000}{1.225 \times \sqrt{9.81 \times 2.0}} \right]^{0.61} = 3{,}089.38 \text{ m} \sim 3{,}090 \text{ m}$$

Example 1.2

A pool fire scenario arises from a leak in a vessel containing gasoline. About 3 gallons of gasoline spilled along a wall with an area of spread of about 1 m². A cable tray, housing pipelines, and electric cables are located 2.5 m above the floor (fire source). Compute the height of the wall flame of fire and also check whether the cable tray will be damaged due to this fire outbreak.

Data
Mass burning rate of gasoline = 0.055 kg/m²-s
Effective heat of combustion of gasoline, ΔH_c = 43,700 kJ/kg

To calculate the diameter of the pool
Area of the dike = 1 m²

Diameter of the dike, $D_{dike} = \sqrt{4 \dfrac{A_{dike}}{\pi}} = 1.128\,\text{m} = D_{pool}$

To calculate HRR
Empirical constant = 2.1
Using Eq. (1.2), $\dot{Q} = (0.055)(43,700)\left[1 - e^{-2.1 \times 1.128}\right](1.0) = 2,178.55\ \text{kW}$
We also know that 1 Btu/h = 0.000293 kW
1 Btu/s = $(3,600) \times (0.000293) = 1.0548\,\text{kW}$
Hence, HRR = 2,065.37 Btu/s

To calculate the length of fire (*L*)
$L * W = A_{dike}$ (As the fire is abutting against the wall, it will act as a dike)
Taking width of fire as 1.2 m, $L = (1/1.2) = 0.83\,\text{m}$

To calculate HRR per unit length of fire

$$\dot{Q} = \frac{\dot{Q}}{L} = \frac{2,178.55}{0.83} = 2,624.76\ \text{kW/m}$$

To compute the flame height of wall fire
Using Eq. (1.4),

$$H_{f,\,wall\ fire} = 0.034\,(2,624.76)^{\frac{2}{3}} = 6.487\,\text{m}$$

As the cable tray is located at 2.5 m above the fire source, it will be damaged.

Example 1.3

A pool fire scenario arises from a leak in a vessel containing hydrocarbon. About 15 gallons of the fuel spilled to form a line along the wall of about 2.5 m². A cable tray, housing pipelines, and electric cables are located at 3 m above the floor (fire source). Compute the height of the wall flame of fire and also check whether the cable tray will be damaged due to this fire outbreak. Take the hydrocarbons as gasoline, diesel, and methanol.

Data:

Fuel	Mass Burning Rate	ΔH_c	$k\beta$
Diesel	0.045	44,400	2.1
Methanol	0.017	20,000	100
Gasoline	0.055	43,700	2.1

To calculate diameter of the pool
Area of the dike $= 2.5\,\text{m}^2$

Diameter of the dike, $D_{dike} = \sqrt{4\,\dfrac{A_{dike}}{\pi}} = 1.784\,\text{m} = D_{pool}$

(a) Gasoline

To calculate HRR
Using Eq. (1.2), $\dot{Q} = (0.055)(43,700)\left[1 - e^{-2.1 \times 1.784}\right](2.5) = 5,866.93\,\text{kW}$
Hence, HRR $= 5,562.12$ Btu/s

To calculate the length of fire (L)
$L * W = A_{dike}$ (As the fire is abutting against the wall, it will act as a dike)
Taking width of fire as $1.5\,\text{m}$, $L = (2.5/1.5) = 1.67\,\text{m}$

To calculate HRR per unit length of fire

$$\dot{Q} = \frac{\dot{Q}}{L} = \frac{5,866.93}{1.67} = 3,513.13\,\text{kW/m}$$

To compute flame height of wall fire
Using Eq. (1.5),

$$H_{f,\ wall\ fire} = \frac{0.034}{2}(3,513.13)^{\frac{2}{3}} = 3.94\,\text{m}$$

As the cable tray is located at 3 m above the fire source, it will be damaged.
(b) Diesel

To calculate HRR
Using Eq. (1.2), $\dot{Q} = (0.045)(44,400)\left[1 - e^{-2.1 \times 1.784}\right](2.5) = 4,877.11\,\text{kW}$
Hence, HRR $= 4,623.73$ Btu/s

To calculate the length of fire (L)
$L * W = A_{dike}$ (As the fire is abutting against the wall, it will act as a dike)
Taking width of fire as $1.5\,\text{m}$, $L = (2.5/1.5) = 1.67\,\text{m}$

To calculate HRR per unit length of fire

$$\dot{Q} = \frac{\dot{Q}}{L} = \frac{4,877.11}{1.67} = 2,920.43\,\text{kW/m}$$

To compute flame height of wall fire
Using Eq. (1.5),

$$H_{f,\ wall\ fire} = \frac{0.034}{2}(2,920.43)^{\frac{2}{3}} = 3.48\ \text{m}$$

As the cable tray is located at 3 m above the fire source, it will be damaged.
(c) Methanol

To calculate HRR
Using Eq. (1.2), $\dot{Q} = (0.017)(20,000)\left[1 - e^{-2.1 \times 1.784}\right](2.5) = 850\,\text{kW}$
Hence, HRR $= 805.84$ Btu/s

To calculate the length of fire (L)
$L * W = A_{dike}$ (As the fire is abutting against the wall, it will act as a dike)
Taking width of fire as $1.5\,\text{m}$, $L = (2.5/1.5) = 1.67\,\text{m}$

To calculate HRR per unit length of fire

$$\dot{Q} = \frac{\dot{Q}}{L} = \frac{850}{1.67} = 508.98 \text{ kW/m}$$

To compute flame height of wall fire
Using Eq. (1.5),

$$H_{f,\,wall\,fire} = \frac{0.034}{2}(508.98)^{\frac{2}{3}} = 1.09 \text{ m}$$

As the cable tray is located at 3 m above the fire source, it will not be damaged.

1.3 JET FIRE

Combustion of fuel on a continuous basis causes jet fire, which has a significant momentum to propagate in the downwind direction. It is a turbulent diffusion flame, continuously released in a predominant direction. Any building modules and equipment, even located far away from the source of jet fire, can be affected seriously. Jet fire releases gases while propagates forward, along and across the direction of propagation. Jet fire may cause damage to the structural members by degrading their strength and functional use. Storage vessels and tanks, located on the downwind side may also be affected, resulting in failure of pipe lines and other supply cables (Chandrasekaran and Kiran, 2015). Depending upon the type of the fuel under combustion, jet fire may release heat flux in the range (200–400 kW/m^2). The properties of jet fire depend on the fuel composition, release conditions, release rate, release geometry, direction, and ambient wind conditions (Chandrasekaran, 2010). While the low-velocity, two-phase releases of jet fire can result in high-radiative flames, similar to pool fire, sonic release of natural gas can produce a high-velocity fire that are less radiative. One of the potential sources of a jet fire is the pressurized gas pipelines. In case of a leak, the initial gas release rate is given by:

$$Q_0 = C_D\, A\, p_0 \sqrt{\left(\frac{MV}{RT_0}\right)\left(\frac{2}{\gamma+1}\right)^{\gamma^2-1}} \tag{1.7}$$

The above equation is valid only if the operational pressure of the gas is higher than the absolute pressure and is related as below:

$$p_0 > p_a\left(\frac{2}{\gamma+1}\right)^{\left(\frac{\gamma-1}{\gamma}\right)} \tag{1.8}$$

where C_D is the discharge (or leakage) coefficient, A is the area of the rupture or the outlet (m^2), p_0 is the operational pressure of the gas, p_a is the absolute pressure, M is the molecular weight of gas (g/mol), V is the rate of specific heat, R is the universal gas constant ($= 8.314$ J/kg/K), γ is the ratio of heat capacities (C_p/C_v), (C_p, C_v) are heat capacities under constant pressure and volume, respectively, and T_0 is the operational temperature (Kelvin). Due to the momentum of jet release, there will be an uplift of jet fire away from the release point. The lift-off distance can be computed using Chamberlain model (Chamberlain, 1987). The following equation is useful to determine the flame length of the jet fire:

$$m = 11.14(Q_o)^{0.447} \tag{1.9}$$

where Q_o is the initial release rate (kg/s). The release rate of the jet fire is given by the following relationship:

$$M_g = \left[\frac{p_0 \ M}{0.08314} \right] \pi r^2 \ L \tag{1.10}$$

$$Q_t = Q_0 \ e^{\left(\frac{Q_0}{M_g} \right) t} \tag{1.11}$$

where M_g is the mass of the gas (kg), r and L are the radius and the length of the pipe (m), respectively, and t is the time of release (s). In case of a leak of combustible material from a pressurized pipeline, a jet flow can occur resulting in jet fire (Tong Shu Jiao et al., 2013). In such cases, the equivalent jet diameter can be computed based on the rupture cross section, as given below:

$$D_{eq} = D_0 \ \sqrt{\frac{\rho_0}{\rho}} \tag{1.12}$$

where D_0 and D_{eq} are the diameter (m) of the outlet (rupture) and the equivalent jet, respectively. ρ_0 and ρ are the density of the leaking gas and the atmospheric gas (kg/m³), respectively. It can be easily seen from the above equation that for the same density of the leaking gas and the atmospheric gas, diameter of the jet will be same as that of the leakage vent/fault. It is necessary to then determine the gas concentration of the released gas from the origin in the direction of propagation of the jet fire, which is given by the following relationship:

$$C(x) = \frac{\dfrac{b_1 + b_2}{b_1}}{0.32 \left(\dfrac{x}{D_{eq}} \right) \sqrt{\dfrac{\rho}{\rho_0}} + 1 - \rho} \tag{1.13}$$

While the above equation helps determine the gas concentration at (x) from the source, it depends on the distribution parameters and are given by the following empirical relationships:

$$b_1 = 50.5 + 48.2 \ \rho - 9.95 \ \rho^2 \tag{1.14}$$

$$b_2 = 23.0 + 41.0 \ \rho \tag{1.15}$$

Further, the concentration of the released jet in any point (y) on a plane, normal to the jet axis is given as below:

$$C(x,y) = C(x)\, e^{-b2\left(\frac{y}{x}\right)^2}$$ (1.16)

As the above equation indicates an exponential decay, the jet concentration reduces with the increase in the distance from the origin of leak, until the jet velocity becomes equal to that of the surrounding wind velocity. In general, the velocity of jet fire, emerging from the gas leakage, is given by the following empirical relationship (Tong Shu Jiao et al., 2013):

$$U(x) = U_0 \left(\frac{\rho_0}{\rho}\right) \left(\frac{b_1}{4}\right) 0.32 \left(\frac{x}{D_{eq}}\right) \left(\frac{\rho}{\rho_0} + 1 - \rho\right) \left(\frac{D_{eq}}{x}\right)^2$$ (1.17)

$$U_0 = \frac{Q_0}{C_d\, \rho\, \pi \left(\frac{D_0}{2}\right)^2}$$ (1.18)

where U_0 is the initial velocity of the jet flow at the leakage point. The leakage coefficient (C_d) lies in the range (0.61, 1). The radiation flux (watts) of each point within the heat source is given by:

$$q = \beta\, Q_0\, H_c$$ (1.19)

where β is the efficiency factor and H_c is the combustion heat of the leaking gas (J/kg). The flame length of the jet fire in such cases can be estimated using the following relationship:

$$L = \frac{(Q_0\, H_c)^{0.444}}{161.66}$$ (1.20)

The radiation strength of the jet fire at any distance (x) away from a point of leakage, measured along the axis of jet flow is given as follows:

$$I = \frac{\alpha\, q}{4\, \pi\, x^2}$$ (1.21)

where α is the parameter accounting for the radiation rate (= 0.2 under normal conditions). In case of multiple sources of leakage, total radiation strength can be computed as the sum of the radiation strength, emitting from all the individual sources of leakage.

1.4 FIREBALL

An immediate ignition of any flammable material released due to the failure of a closed container under over pressure caused by an external heat source; such a failure is also referred to as BLEVE (Boiling Liquid Expanding Vapor Explosion). Major hazards of the fireball are thermal radiation and direct flame contact. Fireball diameter, combustion period, and height at which the fireball shall be formed can be estimated using the following empirical relationships (CCPS, 1994, 1989):

$$D_{max} = 5.8 \sqrt[3]{M} \tag{1.22}$$

$$D_{initial} = 1.3 \, D_{max} \tag{1.23}$$

$$T_{fireball} = 0.45 \sqrt[3]{M} \quad \text{for } M < 30 \text{ tons} \tag{1.24}$$

$$T_{fireball} = 0.26 \sqrt[6]{M} \quad \text{for } M > 30 \text{ tons} \tag{1.25}$$

$$H_{fireball} = 0.75 \, D_{max} \tag{1.26}$$

where M is the initial mass of the flammable liquid (kg). Fireball emerges thermal radiation, whose flux, as received by a black body receptor (energy $A^{-1}T^{-1}$), is given below:

$$E_\tau = \tau_a \, E \, F_{21} \tag{1.27}$$

$$\tau_a = 2.02 \left(p_w \, X_s \right)^{-0.09} \tag{1.28}$$

$$E = \frac{R \, M \, H_c}{\pi \, D_{max}^2 \, T_{fireball}} \tag{1.29}$$

where E_τ is the emissive radiative flux, τ_a is the atmospheric transmissivity, E is the surface-emitted radiative flux, F_{21} is a view factor, p_w is the partial pressure of water, X_s is the distance from the flame surface to the target, H_c is the combustion heat of the leaking gas (J/kg), and R is the radiative fraction of the heat combustion. It is taken as 0.3 and 0.4 for fireballs from vessels below the relief set pressure and the rest of the cases, respectively.

Atmospheric transmittance is the capacity of the atmosphere to transmit electro-magnetic energy, which varies for radiations of different wavelengths; the smaller the atmospheric absorption and scattering of light, the greater the atmospheric trans-mittance. The average temperature of the Earth's surface is about 288 K. For the single-layer atmosphere model, this temperature would correspond to an atmospheric emissivity of about 0.8. Under a BLEVE release, the flammable which may also ignite and cause an explosion and thermal radiation hazards. The duration of the heat pulse under BLEVE is about 10–20 s, causing high-potential damage. The maximum emissive power that results from BLEVE is about 270–333 kW/m^2 in the upward/downward wind and 278–413 kW/m^2 in the crosswind.

1.5 FLASH FIRE

It is a non-explosive combustion of a vapor cloud resulting from a release of any flammable material into open air. The vapor cloud explodes only under certain conditions in areas where intensely turbulent combustion is developed. Major hazards from a flash fire are thermal radiation and direct flame contact. The radiation effect of flash fire is computed similar to that of a fireball. Flash fire results after a substantial delay between the release of flammable materials and the subsequent ignition. It initially forms a vapor cloud over a larger area and then expand radially. It is characterized by a wall of flame. Similar to fireballs, flash fire can also ignite and remain as a continuous flame, which sustains for a longer duration. The instantaneous effect causes thermal radiation, and the flash fire generates 'knock-on' events such as pool fire, jet fire, and BLEVE. It is important to note that the severity of the flash fire is extremely high.

1.6 EXPLOSION

When a vessel containing a pressurized gas rupture, the stored energy is released. If the contents are flammable, then the ignition of the released gas could cause more consequences. Therefore, an explosion is a rapid expansion in volume, resulting in an outward release of energy. The energy is released at high temperatures and pressures. Supersonic explosions created by high explosives are known as detonations and travel through shock waves. Subsonic explosions are created by low explosives through a slower combustion process known as deflagration. Chemical explosives are the most common artificial explosives involving a rapid and violent oxidation reaction that produces large amounts of hot gas. Such explosions, which usually occur in fuel tanks, are initiated by an electric spark or a flame in the presence of oxygen. Explosion behavior is influenced by many parameters, namely ambient temperature and pressure, explosive material composition and its physical properties, nature of ignition source, duration of ignition, amount and turbulence of the explosive material, time before ignition, and the release rate of the explosive material (Crowl and Joseph, 2002).

To understand the consequences of explosion impacts, one need to understand the dynamics of pressure waves. They propagate in air, termed as blast waves, which is followed a strong wind. A shock front is expected from highly explosive materials such as trinitrotoluene (TNT), which generally occurs as a result of sudden rupture of a pressure vessel. In such cases, the maximum pressure above the ambient pressure is generated and termed as peak overpressure. It is very important to understand how this pressure variation is quantified, as the blast wave passes. If it is measured normal to the blast wave propagation direction, then it is termed as side-on overpressure, also known as free-field overpressure. If it is measured facing the incoming shock waves, then it is termed as reflected overpressure. The latter is generally twice in magnitude of the former and can reach as high as even eight times. Explosions result in a blast or a pressure wave moving out from explosion center at the speed of sound. Shock wave or overpressure is the basic cause for damage. Missiles and projectiles are other important sources of damage. Damage is a function of the rate of pressure increase and the duration of blast wave. Blast damage is estimated based on the peak side-on overpressure.

Figure 1.1 shows the variation of peak over pressure with time. Area under this curve is the measure of severity of explosion. As seen in the figure, explosion generates a rapid rise in pressure while the propagating wave causes damage to the objects located on its path. It is then followed by a negative pressure wave which causes further damage before the pressure returns to the atmospheric state. Thus, damage depends on the maximum pressure reached, velocity of propagation, and environmental characteristics.

Explosive force is released in a direction normal to the surface of the explosive, but if released in midair, then the direction of blast will be 360°. The speed of an explosive reaction is very rapid, resulting in the thermal expansion of gases causing high pressure (Dubnikova et al., 2005). The generation of heat in large quantities accompanies most explosive chemical reactions. The rapid liberation of heat causes instantaneous expansion of gases, generating high pressure. It is important to note that the liberation of heat at a slower pace will not cause an explosion. A substance that burns less rapidly (slow combustion) evolves more heat in comparison to that detonates rapidly (fast combustion). In the former, slow combustion converts more of the internal energy of the burning substance into heat released to the surroundings. In the latter, fast combustion converts more internal energy into work on the surroundings. A chemical explosive, under heat or shock, decomposes very rapidly and yields gas and heat. A explosive reaction must be capable of being initiated by the application of shock, heat, or a catalyst to a small portion of the mass of the explosive material.

Rupture of a pressurized vessel resulting in an explosion could occur due to the following reasons, namely failure of a pressure regulator or any similar pressure-relief component, reduction in the vessel thickness due to corrosion, erosion or any similar chemical attack, reduction in the vessel strength due to overheating, chemical attack,

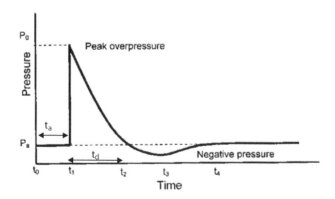

FIGURE 1.1   Variation of overpressure with time. (Courtesy: Chandrasekaran, 2016.)

fatigue-induced weakening, internal runaway reaction, or any other incident result-ing in loss of process containment. Under explosion, the amount of energy released will result in the following, namely vessel stretching or tearing, kinetic energy of the fragment, and energy in shock wave or as a waste energy, released in the surrounding air. Major human effects caused by fire and explosion are summarized in Table 1.2 (Assael and Kakosimos, 2010).

Studies also showed that the overpressure that resulted from the explosion is not significant in open areas. Further, a complex geometric layout of a process plant or an offshore drilling unit can initiate high explosion over pressure (Dadashzadeh et al., 2013).

Explosions in process industries and offshore platforms lead to severe conse-quences. Recent studies showed that as much as two-thirds of the losses arising from an accident in such industries are due to explosions; more than three-quarters of the explosion involve combustion or explosive materials. Table 1.3 summarizes the main causes, frequent locations of accidents in such plants, and various contribut-ing factors (Lees, 1996). It is evident that about 18% of fire are due to release and overflow of flammable gases and/or liquids. Fire contributes about 20% to the total loss, while the rest of the major contribution comes from the explosions (Faisal and Abbassi, 1999). Fire and explosion accidents in maritime industry arise from four major groups of activities, namely human error, mechanical failure of plants and equipment, thermal reaction, and electric faults (Baalisampang et al., 2018); their percentage contributions are reported as (48, 22, 14, 7), respectively. The major fac-tors that contribute to fire and explosions due to human error are poor maintenance scheduling, inadequate training and education, and lack of situation awareness. Other factors that contribute to mechanical faults are error in maintenance works, metal fatigue, excess heat generation, and faulty operational conditions. Short circuit, poor connection and maintenance, and use of defective devices are major reasons for electric faults causing fire and explosions.

TABLE 1.2
Human Effects Caused by Explosion

Effects of Explosion	Damage
Probability of eardrum injury	Ear drum rupture is a direct effect of overpressure during an explosion
Probability of death due to lung damage	Explosion can cause a sudden pressure difference between the inside and the outside of the lungs due to which the thorax is pressed inward, causing lung damage and possibly death
Probability of death from head impact	Shock wave can push the head of a person backward, resulting in skull rupture or fracture or even the collision of the head with another stationary or non-stationary object
Probability of death due to whole body displacement	Shock wave can throw the whole body backward, causing death because of its impact with other objects

TABLE 1.3

Cause and Factors of Explosion in Process Industries

Description	Proportion
Main cause	
Chemical reaction uncontrolled	20.0
Chemical reaction accidental	15.0
Combustion explosion in equipment	13.3
Unconfined vapor cloud	10.0
Overpressure	8.3
Decomposition	5.0
Combustion sparks	5.0
Pressure vessel failure	3.3
Improper operation	3.3
Others	16.8
Frequent location of occurrence	
Enclosed process or manufacturing buildings	46.7
Outdoor structures	31.7
Yard	6.7
Tank farm	3.3
Boiler house	3.3
Others	8.3
Contributing factors	
Rupture of equipment	26.7
Human element	18.3
Improper procedures	18.3
Faulty design	11.7
Vapor-laden atmosphere	11.7
Congestion	11.7
Flammable liquids	8.3
Long replacement time	6.7
Inadequate combustion controls	5.0
Inadequate explosion relief	5.0

1.7   EXPLOSION CLASSIFICATIONS

Explosions can be classified into four basic types, namely Vapor Cloud Explosion (VCE), Pressure Vessel Explosion (PVE), Boiling Liquid Expanding Vapor Explosion (BLEVE), Condensed-Phase Explosion (CPE), and Dust Explosion (DE).

1.7.1   VAPOR CLOUD EXPLOSION (VCE)

The following conditions are necessary for a VPE:

1. There must be a release of flammable material at an appropriate temperature and pressure. The flammable materials include liquefied gases under pressure, ordinary flammable liquids, and gases. When a flammable liquid spills, it vaporizes, and termed as vapor cloud.

2. Ignition must be sufficiently delayed for a vapor cloud to form a maximum flammable cloud size. It usually takes about 60 s so that the ignition delay is not too long. If the ignition occurs instantly, it will result in a fireball.
3. The fuel-air ratio of a sufficient amount of the vapor cloud must be present within the flammable range. A uniform fuel-air mixture closer to the stoichiometric fuel-air ratio will cause a stronger explosion.
4. There must be a flame-acceleration mechanism to be present for causing VCE; congested areas within the flammable portion of the vapor cloud are good examples.

The speed of the flame propagation governs the overpressure produced by a VCE. Objects that are present in the pathway of the flame enhance the turbulence of both the vapor and the flame. Formation of turbulence further enhances the speed of the flame, which causes overpressure. Confinement of the space that limits flame explosion also increases the speed of flame propagation. If the flame acceleration is controlled, one can avoid explosion, but will result in a large fireball or flash fire. Thus, the center of VCE is not necessarily where the flammable material is released, but the congested area within the vapor cloud. Kindly note that work-floors with multiple areas of congestion can cause series of explosions as the flame shall propagate through each of these congested areas. A better design for explosion resistance would be to plan more open areas without congestion.

1.7.2 PRESSURE VESSEL EXPLOSION (PVE)

PVE may occur as anyone (or their combination) of the following types:

- Deflagrations and detonations of pure gases, which are not mixed with the oxidants. Example: Acetylene.
- Combustion deflagrations and detonations in the enclosures, which can occur in the presence of gaseous, liquid, or dust particle fuels. If an enclosure is too weak to sustain the pressure resulting from such combustion, it will result in PVE.
- Runaway exothermic chemical reactions can cause an accelerated condition if there is any delay in the process of removing the released energy. If the pressure in the containment vessel exceeds the pressure capabilities of the vessel, it will result in PVE.
- An overpressure of the equipment with nonreactive gaseous contents can also result in PVE, but termed as mechanical explosion (ME).
- When two streams of a mixture with a widely different temperature occur suddenly, it may result in PVE. It can cause flashing of the cooler liquid to generate vapor and develop pressure. If this developed pressure exceeds the vessel capacity, it may result in PVE.

1.7.3   BOILING LIQUID EXPANDING VAPOR EXPLOSION (BLEVE)

BLEVE is a special form of PVE, which occurs when a large amount of pressurized liquid is suddenly vented into the atmosphere. BLEVE may cause a huge billowing and emit radiant fireballs. The primary cause of BLEVE is impingement of an external flame on the vessel, above the liquid level. It weakens the container and causes rupture. BLEVE also occurs when a sealed container of liquefied gases such as LPG is accidentally exposed and enveloped by fire. Vapor is generated, causing a rapid rise in internal pressure. As a result, the container wall temperature rises, and thus weakens its strength. Even in the presence of a pressure-relief valve (PRV), stress imparted by this increased pressure is more than the wall strength; it causes rupture of the vessel, releasing the superheated liquid. The released superheated liquid expands and vaporizes within a very short time (milliseconds), causing catastrophic damage from the spread of the ignited vapor. Figure 1.2 shows the flow chart of BLEVE.

When a pressure-containing liquid above its atmospheric boiling point ruptures, BLEVE can occur due to the rapid vaporization. If it is flammable, then a delayed ignition could cause either a VCE or a flash fire while instantaneous ignition would result in a fireball. Under such situations, field professionals are often confused whether BLEVE could result from the opening of the pressure-relief valve (PRV). Please note that BLEVE can result neither from the opening of PRV nor PRV can be used as a device to protect against BLEVE. Common situations that result in BLEVE are mechanical damage, caused by corrosion or collision, overfilling of the vessel without any check valve or relief valve, overheating of the pressure vessel with an inoperative relief valve, vapor space explosion, or a hazard situation of exposure to fire; the last one is the most common reason for BLEVE in process plants.

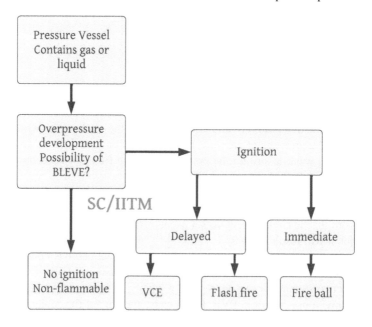

FIGURE 1.2   Flowchart of BLEVE.

BLEVE can cause a major accident which can have severe consequences. Overpressure and ejection of vessel fragments are the common effects of such an explosion, followed by a fireball if the substance is flammable (Reid, 1979). When a vessel undergoes a BLEVE, part of the released mechanical energy is converted into overpressure. There are different methodologies to calculate this mechanical energy, based on the diverse thermodynamic assumptions (Planas and Casal, 2015):

- Constant volume energy addition (Brode, 1959)
- Real gas behavior and isentropic expansion (CCPS, 2010)
- Isothermal expansion (Smith et al., 1996)
- Thermodynamic availability (Crowl, 1991, 1992a, b)
- Ideal gas behavior and isentropic expansion (Prugh, 1991a, b; Carl and William, 2001)
- Real gas behavior and adiabatic irreversible expansion (Planas et al., 2004; Casal and Salla, 2006)

Under exposure to fire below the liquid level in the pressure vessel, heat of vaporization of the liquid develops a heat sink, which is similar to a teakettle. Under normal conditions, the evolved vapor will exit through the relief valve. But, in case the flame impinges on the vessel above the liquid level, walls of the vessel are weakened causing rupture of the vessel, even if the relief valve is opened. Superheat provides the explosive energy for BLEVE, which is the maximum at the superheat temperature. As understood from the above discussion, BLEVE can occur only due to mechanisms that cause sudden rupture or failure of the vessel (or container). A sudden rupture normally occurs under impact of a sharp object, corrosion, manufacturing defects, or internal overheating.

Accidents caused by BLEVE have serious consequences. See, for example, the Mexico (1984) accident at St. Ixhuatepec where BLEVE explosion of a storage tank took place. Figure 1.3 shows the exploded tank; location and details are masked due to strategic reasons. This tank contained LPG of about 12,000 m^3, and this was one of the largest LPG gas supplies to the city. The explosion caused more than 500 deaths and about 5,000 injured and referred to as one of the deadliest industrial disasters in the world. Gas leak initiated the disaster, likely caused by a pipe rupture during transfer operations. The plume of LPG was concentrated at ground level for about 10–12 min, which subsequently grew much larger and was drifted toward windward side. The explosion shock wave destroyed dwelling units located up to 7 km range, while window glass panes were found to be damaged in houses located event at a distance of about 10 km. Radiant heat generated by the inferno incinerated most corpses to ashes, with only 2% of the recovered remains left in recognizable condition (Arturson, 1987).

Blast effects from BLEVE are insignificant, but important in the near-field design. One of the serious consequences that influences the design is the large quantity of TNT-equivalent release. The blast wave thus produced depends on a few factors, namely type of the fluid released, energy during release, rate of energy release, shape and size of the pressure vessel, and the type of rupture. BLEVE effects consider the rapid expansion (flashing) of the liquid and the expansion of the compressed vapor

FIGURE 1.3   LPG tank that suffered BLEVE.

present in the head space of the pressure vessel. The explosion energy arising from BLEVE is given by:

$$E = \frac{(P_1 - P_2)\, V}{\gamma - 1} \qquad (1.30)$$

where E is the explosion energy, V is the volume of the vessel under consideration, and (P_1, P_2) are the initial and final pressures. Exergy concept is used to calculate the energy released during BLEVE (Crowl, 1992). Exergy, referred to as thermodynamic availability, is the maximum potential work that can be performed by a system when it reaches to the ambient state in terms of thermal, mechanical, and chemical equilibrium.

For a BLEVE, this can be interpreted as the superheated liquid flashes from its initial state to its final state, where the final state is defined by the ambient temperature and pressure. Using the analogy of maximum potential work done by a compressed gas or a combustible liquid, batch energy released during BLEVE is given by (Russell et al., 2011):

$$E = (U - U_0) + p_0(V - V_0) - T_0\,(S - S_0) \qquad (1.31)$$

where the subscript zero denotes the dead state (which used to be the ambient temperature and pressure and is usually taken as (298 K, 101.3 kPa)), E is the batch energy (kJ/kg), U is the internal energy (kJ/kg), p is the absolute pressure (N/m²), V is the specific volume (m³/kg), T is the absolute temperature (Kelvins), and S is the entropy (kJ/K kg).

Example 1.4

A vessel containing propane to its full capacity is assumed to BLEVE at 90% of its critical temperature. The initial pressure based on the saturation pressure at the critical temperature is 2,104 kPa. For the saturated liquid, internal energy, entropy, and specific volume are obtained from the experiments. Propane expands into the natural environment, setting a dead state property as (1 bar, 25°C). Calculate the energy released during BLEVE. Use the data given below:

Critical temperature of propane = 369.8 K
Internal energy at critical temperature and dead state = (258.9, 471.1) kJ/kg
Specific volume at critical temperature and dead state = (2.34 × 10⁻³, 0.55)
m³/kg
Entropy at critical temperature and dead state = (0.9, 2.22) kJ/K Kg).

Solution

$$E = (U - U_0) + p_0 (V - V_0) - T_0 (S - S_0)$$

$$E = (258.9 - 471.1) + 101.3 (2.34 \times 10^{-3} - 0.55) - 298 (0.9 - 2.22) = 125.68 \text{ kJ/kg}$$

This is the maximum (thermochemical) exergy released by the propane system under BLEVE; lesser values are possible with lesser values of superheat. Table 1.4 is useful for referring to thermodynamic properties of a few chemicals.

Example 1.5

A cylindrical vessel of volume 50 m³ contains liquid propane at room temperature (25°C). It undergoes BLEVE due to fire engulfment, resulting in the loss of containment through a safety valve. At the time of explosion, temperature was 50°C and the filling degree is 25°. Estimate the scaled overpressure at a distance of 125 m.

TABLE 1.4
Thermodynamic Properties of Selected Chemicals

Chemical	Normal Boiling Point (K)	Critical Temperature (K)	Critical Pressure (MPa)
Propane	231.1	369.8	4.25
Chlorine	239.2	416.9	7.98
Ammonia	239.8	405.5	11.35
Water	373.2	647.3	22.12

Data given:

Filling degree, FD = 0.25 (expressed in %100)
$T = 50\,°C = 323$ K

Solution

Mechanical energy (e), released per m³ of a vessel, as a function of explosion temperature and initial filling degree (expressed in parts per unit, instead of percentage) is given the following empirical relationship (Hemmatian et al., 2015, 2017):

$$e = 43.97 - 213.9\ FD - 0.152\ T + 1.349\ FD\ (T) - 4.36 \times 10^{-4}\ T^2 - 2.05 \times 10^{-3}$$

$$FD\ (T^2) + 1.55 \times 10^{-6}\ T^3 \tag{1.32}$$

$$e = 43.97 - 213.9\ (0.25) - 0.152\ (323) + 1.349\ (0.25)(323) - 4.36 \times 10^{-4}\ (323^2) - 2.05$$

$$\times 10^{-3}\ (0.25)(323^2) + 1.55 \times 10^{-6} \times (323^3) = 3.738\ MJ/m^3$$

Total energy released, $E = 3.738 \times 50 = 186.9$ MJ.

TNT-equivalent mass can be computed using the following relationship (Chandrasekaran, 2016):

$$m_{TNT} = \frac{\eta E}{E_{TNT}} \tag{1.33}$$

where η is the energy of explosion (usually varies between 1% and 15%), E is the total released energy, and E_{TNT} is the energy of explosion of TNT, which is about 4,686 kJ/kg:

$$m_{TNT} = \frac{0.15 \times 186.9 \times 10^3}{4,686} = 5.98\ \text{kg TNT} \tag{1.34}$$

The scaled distance can be computed using the following relationship (Chandrasekaran, 2016):

$$Z_e = \frac{r}{\sqrt[3]{m_{TNT}}} = \frac{125}{\sqrt[3]{5.98}} = 68.87 \tag{1.35}$$

Overpressure can be estimated using the scaled overpressure curve (Chandrasekaran, 2016).

Scaled overpressure at 68.87 m is found to be 0.04 bar (4 kPa), using Figure 1.4. The following Matlab code helps to plot the curve.

This program plots scaled distance to scaled overpressure
Program code written by Pradeepa and Chandrasekaran

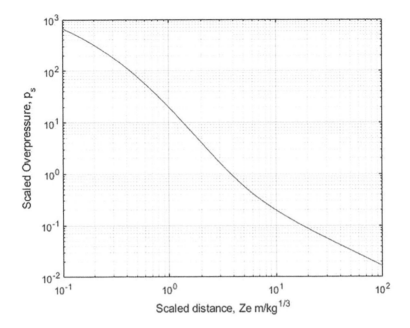

FIGURE 1.4   Scaled overpressure. (Courtesy: Chandrasekaran, 2016.)

```
ze=logspace(-1,2);
y1=1616*(1+(ze/4.5).^2);
y2=1+(ze/0.048).^2;
y3=1+(ze/0.32).^2;
y4=1+(ze/1.35).^2;
y5=sqrt(y2.*y3.*y4);
Y=y1./y5;
loglog(ze, Y)
grid on
xlabel('Scaled distance, Ze m/kg^{1/3}')
ylabel('Scaled Overpressure, p_s')
```

As in case of oil exploration and production platforms where hazardous gas and petrochemicals are processed, a BLEVE is quite likely to occur when such platforms are exposed to fire (Tellez and Pena, 2002). The possibility of BLEVE can be assessed using Reid's massive nucleation theory (Reid, 1979). The zone of spontaneous nucleation of a vulnerable petrochemical can be easily understood using pressure-temperature diagram, shown in Figure 1.5. As seen in the figure, a tangent line is drawn from the critical (p, T) point to a point where the ordinate represents atmospheric pressure. A horizontal line is drawn from a typical pressure value to intersect the curve. From the point of intersection, a vertical line is extended as shown. The enclosed area as hatched in the figure indicates the spontaneous nucleation of the chosen petrochemical. This is the suitable scenario for BLEVE to occur. It is clear that when the vessel is exposed to fire, the released heat increases the temperature and the corresponding pressure inside the vessel. This progressive heating will lead

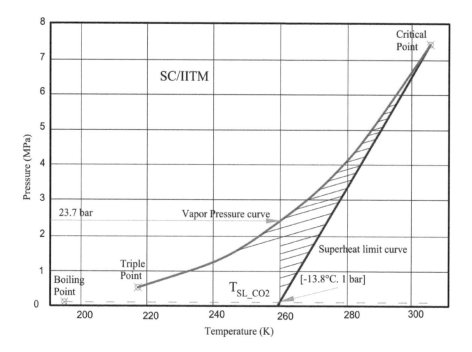

FIGURE 1.5 Pressure-temperature curve indicating spontaneous nucleation.

to a point where the vertical line is bypassed. Under such a condition, a sudden rupture is likely to occur and termed as BLEVE.

Effects of BLEVE are of three types, namely thermal radiation, overpressure effect in the shock waves, and projection of the fragments. The thermal radiation effects are attributed to the fireball (see Figure 1.2). Empirical relationships used to compute the maximum diameter of the fireball, height of the center of fireball, and time taken to form a fireball can be seen in Section 1.3. The flow of radiation per unit emissive surface area (kW/m²) is given by CCPS (1989, 1994), Elia (1991), and Pape (1988), respectively:

$$I = \frac{F_R \left(-\Delta H_{comb} \right) M}{\pi \, D_{max}^2 \, t_{BLEVE}} \tag{1.36}$$

$$I = \frac{0.27 \, M \left(-\Delta H_{comb} \right) p_0^{0.32}}{\pi \, D_{max}^2 \, t_{BLEVE}} \tag{1.37}$$

$$I = 235 \, p_v^{0.39} \tag{1.38}$$

where F_R is the ratio between energy emitted by radiation and the total energy released by combustion (value suggested in the literature is (0.25 to 0.4)), ΔH_{comb} is the heat of combustion of the material (kJ/kg), p_0 is the initial pressure at which the flammable chemical is stored (MPa), and p_v is the vapor pressure of the chemical.

The radiation (I_R) received at any distance (X) from the emitting point is computed using the geometric view factor (F_{vg}) and the fraction of energy transferred; the latter is also known as atmospheric transmissivity:

$$\eta = 2.02 \left(p_w \ X \right)^{-0.09} \tag{1.39}$$

$$F_{vg} = \frac{D^2}{4 \ X^2} \tag{1.40}$$

$$I_R = I \ \tau \ F_{vg} \tag{1.41}$$

where p_w is the partial pressure of water at ambient temperature (Pa). Roberts (1982) and Roberts (1952) suggested a different expression for computing the intensity of radiation, which depends only on the mass of the fuel and is given by:

$$I_R = \frac{828 \ M^{0.771}}{X^2} \tag{1.42}$$

BLEVE also causes overpressure effects, which are generally difficult to predict as the duration of rupture de-pressurization is a complex quantity. See Example 1.5 for more details of estimating the overpressure effects. One of the approximate ways to estimate this is through equivalent weight of TNT for a BLEVE event (Prugh, 1991b). TNT is considered as an explosive for two reasons. Primarily it contains elements of carbon, oxygen, and nitrogen, which means that when burns it produces highly stable substances with strong bond. Further, TNT explosions are chemically unstable, which means that it does not require much force to break their bond (Chandrasekaran, 2016).

The prediction of fragment effects is important, as many deaths and domino-damage effects are caused by fragment projectiles. Fragments are not evenly distributed; vessel's axial direction receives more than that is sideways. The total number of fragments is a function of the vessel capacity and is given as follows:

$$\# \text{ of fragments} = -3.77 + 0.0096 \text{ (vessel capacity, in m}^3\text{)} \tag{1.43}$$

The range of validity of the above equation is 700–2,500 m^3.

1.7.4 CONDENSED-PHASE EXPLOSION (CPE)

It will occur when the explosive materials are either in a liquid or a solid phase. Such explosions are more common in organic-based hydrocarbons. Examples are RDX and TNT. Other inorganic compounds based on azide radical readily decompose even under slight heating to become explosives. For example, look at the following chemical reaction:

$$P_b \left(N_3 \right)_2 \rightarrow P_b + 3N_2 \tag{1.44}$$

Also, addition of fulminate radical (for example, mercury fulminate, lead fulminate) shall also cause decomposition to form a ready explosive even under slight impact or friction. These are termed as percussion crackers. Acetylide radicals in the presence of ultraviolet light and slight impact readily decompose to cause explosion.

There also exist inorganic-based condensed explosives such as lead styphnate $P_b \left[C_6 \, HO_2 \, (NO_2)_3 \right]_2$. Hence, an explosive substance can originate either from an organic or inorganic compounds. Condensed-phase systems (solids and liquids) have a high heat of decomposition and capable of detonating. These materials are routinely found in the chemical process industry (AIChE, 1996). Examples include some organic peroxides, acetylenic compounds, and nitration mixtures. Blast and fragment effects from such explosions are evaluated from the energy released in an explosion of an equivalent mass of TNT.

1.7.5 DUST EXPLOSION

Dust is defined as particle whose diameter is lesser than 500 μm. A dust explosion is the rapid combustion of fine particles suspended in the air within an enclosed location. They can occur under the presence of dispersed powdered combustible material in high concentrations in air. If the combustible material is a hydrocarbon fuel, then the explosion is termed as fuel-air explosion. Dust particles, emitted from paper mills, pose a serious dust explosion hazard. One of the simplest methods to control duct explosion is to maintain a relative high level of humidity. Due to the presence of very small particles in dust, it generates a high surface area to volume ratio.

Suspended fine solid particles can explode in a fashion similar to that of the flammable gases. In case of dust suspended in air, even a small concentration of flammable gas can contribute to a severe explosion than that of the presence of the dust particles alone. Such suspended dust particles are referred to as hybrid mixtures. Dust explosions are a frequent hazard in coal mines, grain elevators, and other industrial environments. Many common materials burn to generate a dust explosion, such as coal and sawdust. In addition, powdered metals such as aluminum, magnesium, and titanium can form explosive suspensions in air, if finely divided. A dust explosion can cause major damage to structures, equipment, and personnel due to overpressure or shockwave effects. Intense radiant heat from a fireball can also ignite the surroundings or cause severe skin burns in unprotected persons. In a tightly enclosed space, the sudden depletion of oxygen can cause asphyxiation. In case of coal mines, incomplete combustion of dust may produce large amounts of carbon monoxide, causing more deaths.

Explosive dust arises during transporting grain, mining of coal, and flour mills. One of the classical examples of dust explosion is the dust explosion caused in the Washburn A Mill, Minneapolis on May 2, 1878, killing 18 workers apart from damaging many adjacent buildings. Another major disaster occurred in coal mine is the Benxihu colliery explosion, located in Benxi, Liaoning, Republic of China, on April 26, 1942, causing death of about 1,600 people. This is in fact close to about 35% of the miners who were on duty (Roberts, 1952).

1.8   FIRE AND EXPLOSION CHARACTERISTICS

One of the major differences between fire and explosion is the rate of energy release rate. While the energy release rate is slow in case of fire, it is rapid in explosion; in fact, it is in the order of milliseconds. It is very important to note that fire can result in explosion and an explosion can also cause fire; these events are interchangeable. A few commonly used terminologies in fire and explosion discussed below are useful.

1.8.1   COMBUSTION

Combustion or fire is a chemical reaction. A substance combines with an oxidant and releases energy, while part of the released energy is used to sustain this reaction.

1.8.2   IGNITION

It is referred to as the action of setting anything to fire. Ignition of any flammable mixture occurs when it comes in contact with a source of ignition with sufficient energy to burn. In the case of gases, if the temperature present in the scenario is very high, then it is sufficient to autoignite the gas.

1.8.3   AUTOIGNITION TEMPERATURE (AIT)

AIT of a flammable substance is the lowest temperature at which it ignites spontaneously even in the absence of any ignition source. Under such cases, the temperature alone provides sufficient energy to induce combustion. It is also termed as the ignition temperature of a substance, which is the lowest temperature at which it starts combustion reaction. With the increase in the oxygen content in the atmosphere, the autoignition temperature decreases; it also decreases at a high pressure. The time taken by a flammable material to reach the autoignition temperature is given by (James, 2016):

$$t_{ig} = \frac{\pi}{4} \lambda \rho \, c_p \left[\frac{\left(T_{ig} - T_o \right)}{q''} \right]^2 \tag{1.45}$$

where t_{ig} is the time taken to reach the autoignition temperature, λ is the thermal conductivity of the flammable material, ρ is the density of the combustible material of the material, c_p is the specific heat capacity of the flammable material, $\left(T_{ig}, T_o \right)$ are the autoignition and initial temperature of the material, and q'' is the heat flux density of the material.

1.8.4   SPONTANEOUS COMBUSTION

It occurs due to the increase in temperature caused by the exothermic internal reactions; this initiates self-heating of the material. It is followed by a thermal runaway and autoignition. Spontaneous combustion occurs in materials that possess a relatively low ignition temperature; for example, hay, straw, and peat. These materials release heat either by oxidation in the presence of moisture and air, or by bacterial fermentation. As a result, temperature of the material rises above its ignition point to cause a thermal runaway.

1.8.5 FLASH POINT

Flash point of a volatile chemical (fuel) is the lowest temperature at which it gives off enough vapor to form an ignitable mixture with air (Crowl and Joseph, 2002). Flash point temperature does not depend on the ignition source temperature; usually, the latter is far higher than the flash point temperature. Flash point is one of the descriptive characteristics of the fuel used to distinguish flammable fuels from others. The flash point is an empirical measurement rather than a fundamental physical parameter. Experimentally measured values vary with the equipment and test protocol variations, including temperature ramp rate, time allowed for the sample to equilibrate, sample volume, and whether the sample is stirred or not. Table 1.5 shows flammable properties of a few flammable fuel.

1.8.6 FIRE POINT

It is the lowest temperature at which a vapor above a liquid will continue to burn once ignited. Usually, the fire point temperature of a fuel is higher than its flash point temperature, but, in general, it is about 10°C higher than the flash point.

TABLE 1.5
Flammable Fuel Properties

Fuel	Flash Point	Flammability Limits (% vol)		Autoignition Temperature
		LFL	UFL	
Acetone	Flammable gas	2.6 to 3	12.8 to 13	305°C
Benzene	−11°C	1.2	7.8	560°C
Butane, n-butane	−60°C	1.6	8.4	420–500°C
Carbon monoxide	−191°C	12	75	609°C
Ethanol (70%)	16.6°C	3–3.3	19	363°C
Gasoline (petrol)	−43°C	1.4	7.6	246–280°C
Diesel	>52°C	0.6	7.5	210°C
Kerosene	40–70°C	0.6–0.7	4.9–5	220°C
Biodiesel	130°C			Xxx
Methane (natural gas)	Flammable gas	5	14.3	580°C
Naphthalene	79–87°C	0.9	5.9	540°C
Propane	Flammable gas	2.1	9.5–10.1	480°C
Pentane	−40–−49°C	1.5	7.8	480°C
Octane	13°C	1	7	Xxx

1.8.7 FLAMMABILITY LIMITS

These limits are the range of vapor concentration that could cause combustion on intercepting an ignition source. There are two limits in which the fuel shall catch fire, namely Lower and Upper Flammability Limits (LFL, UFL). Lower limit is expressed as the least concentration (in percentage) of a gas or a vapor in air, which is capable of producing flash fire in the presence of an ignition source like arc, flame, or heat. This is also termed as lower explosive level (LEL). At a concentration in air lesser than LFL, the fuel mixture is termed as a lean mixture and shall not catch fire even in the presence of any ignition source. For example, diesel fuel has LFL as 0.6 (% volume), which means that an explosion cannot occur in the presence of external ignition source if the atmosphere has lesser than 0.6% diesel fuel. Explosimeter (also called as combustible gas detector), shown in Figure 1.6, is a gas detector used to measure the % of combustible gas present in the atmosphere. Upon exceedance of LEL, an alarm is raised by the device, to active the immediate fire safety control measures.

Upper limit is the highest concentration (expressed in %) of a gas or a vapor in air, capable of producing a flash fire in the presence of an external ignition source. Concentrations higher than LFL are termed as too-rich mixture and shall not catch fire even in the presence of an ignition source. It is very important to note that processing operations of fuel mixture shall not be carried out above UFL. Because, any air leak shall immediately bring the fuel mixture in the flammability range. Hence, within the LFL and UFL, a fuel mixture is likely to catch fire in the presence of an external ignition source. The National Fire Protection Association (NFPA) prescribes the flammable limits of a few gases and vapor, whose concentration is given in percentage by volume in air. NFPA classifies liquid and gas as Class 1A, Class 1B, Class IC, Class II, Class IIIA, and Class IIIB based on the flash point and the boiling point (Carl and William, 2001). Table 1.6 summarizes the classification.

1.8.8 LIMITING OXYGEN CONCENTRATION

It is the minimum oxygen concentration (MOC) below which combustion is not possible with any fuel mixture irrespective of its concentration. It can also be expressed as the maximum allowed oxygen concentration of the fuel-air-inert mixture in which an explosion will not occur (Domnina et al., 2006). Expressed as volume

TABLE 1.6
Classification of Gases and Vapor (NFPA)

Description	Flash Point	Boiling Point	NFPA 704 Flammability Rating
Class IA	<23°C	<38°C	4
Class IB	<23°C	>38°C	3
Class IC	>23°C	<38°C	
Class II	>38–60°C<	Not specified	2
Class IIIA	>60–93°C<		
Class IIIB	>93°C		1

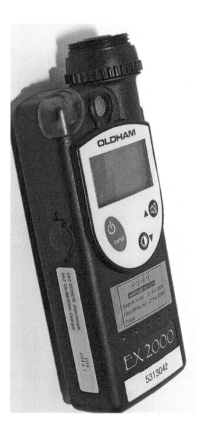

FIGURE 1.6 Explosimeter.

percentage of oxygen, it is also termed as minimum oxygen concentration and maximum safe oxygen concentration (MSOC). The flammability characteristics of materials for their variations with LFL, UFL, and flash point can be seen in Figure 1.7 (Chandrasekaran, 2016).

As seen from Figure 1.7, the flammability limits depend on temperature and pressure. UFL increases with the rise in temperature beyond the ambient temperature while LFL decreases, creating a wide range of flammability region with the rise in temperature. Similarly, UFL increases with the increase in pressure while pressure has no influence on LFL. UFL variation with pressure can be seen from the following relationship:

$$\text{UFL}_{\text{at absolute pressure}} = \text{UFL} + 20.6 \, (\log P \; 1) \tag{1.46}$$

The flammability characteristics of liquid can be determined by Cleveland open-cup method (Wray and Harry, 1992). Knowing the boiling point of the liquid, flash point of the liquid can also be estimated using the following empirical relationship, which is validated with the open-cup tests; the constants (a, b, c), used in the equation are given in Table 1.7 for a few flammable chemicals.

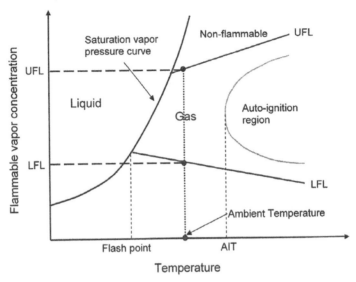

Relationship between various charactertics

FIGURE 1.7   Flammability characteristics of materials. (Chandrasekaran, 2016.)

TABLE 1.7
Flash Point Constants for Flammable Chemicals

Chemical	a	b	c
Hydrocarbon	225.1	537.6	2,217
Alcohol	225.8	390.5	1,780
Amines	222.4	416.6	1,900
Acids	323.2	600.1	2,970
Ethers	275.9	700.0	2,879
Sulfur	238.0	577.9	2,297
Esters	260.5	449.2	2,217
Halogens	262.1	414.0	2,154
Aldehydes	264.5	293.0	1,970

$$T_f \,(\text{in Kelvins}) = a + \frac{b \left(\dfrac{c}{T_b}\right)^2 e^{-\left(\frac{c}{T_b}\right)}}{\left[1 - e^{-\left(\frac{c}{T_b}\right)}\right]^2}$$

(1.47)

1.9 FLAMMABILITY DIAGRAM

Flammability diagrams show the control of flammability in mixtures of fuel, oxygen, and an inert gas, typically nitrogen. Mixtures of the three gases are usually depicted in a triangular diagram, known as a ternary plot. Flammability diagram helps understand whether a given mixture is flammable and also helps to control or prevent fire and explosion of the flammable mixture. A typical flammability diagram is shown in Figure 1.8.

1.9.1 CONSTRUCTING FLAMMABILITY DIAGRAM

As seen in the above, a typical flammability diagram has three arms, namely nitrogen arm (which is the usually the base of the triangle), fuel arm, and oxygen arm, marked in an anticlockwise order. The following steps are useful to construct a flammability diagram.

   Step 1: Mark three arms of the triangle and mark them as nitrogen, fuel, and oxygen in an anticlockwise manner. Keep the nitrogen arm as the base. The origin and the apex of each arm are marked from left to right in the anticlockwise manner as shown in the figure. For example, apex of the nitrogen arm is the origin of the fuel arm and so on.
   Step 2: Mark 79% point on the nitrogen arm, which is the conventional percentage of nitrogen concentration in the atmosphere (rest is oxygen concentration).
   Step 3: Draw a line connecting the apex of the fuel arm to this point of 79% on the nitrogen arm. This is called as the airline, as marked in the figure.

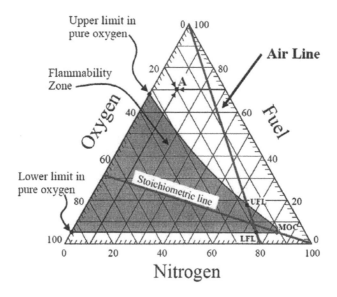

FIGURE 1.8 Flammability diagram. (Chandrasekaran, 2016.)

Step 4: Knowing the percentage of fuel in air, mark the LFL and UFL of the fuel mixture on this airline. Knowing the LFL and UFL values of the fuel in air, mark these limits on the fuel arm and project them on the airline as shown.

Step 5: Mark the stoichiometric point on the oxygen arm. To obtain the stoichiometric point, the equation of the combustion reaction can be used. Let us explain this with an example of methane.

The general equation of combustion is given by the following relationship:

$$C_mH_xO_y + zO_2 \rightarrow mCO_2 + \frac{x}{2} H_2O$$

Balanced combustion reaction for methane is given by:

$$CH_4 + 2O_2 \rightarrow CO_2 + 2H_2O$$

By comparing the above two equations, one can observe the following values, namely ($m = 1, y = 2, x = 4, z = 2$).

Stoichiometric point is given by the following relationship:

$$\frac{z}{(1+z)} \times 100 = 66.7\% \text{ of oxygen}$$

Hence, mark 66.7% on the oxygen arm of the triangle as shown. Draw a line from this stoichiometric point to the apex of the nitrogen arm. This is called as stoichiometric line.

Step 6: Mark the LOC point on the oxygen arm and draw a line parallel to the fuel arm. This line shall intersect the stoichiometric line, and this intersection point is referred to as the nose of the curve.

Step 7: Knowing the LFL and UFL of the fuel in pure oxygen, mark these points on the fuel arm and project them on to the oxygen arm.

Step 8: Connect the points till the nose to form the shaded region, which is the flammable region.

Once the flammable region is established for a particular fuel mixture, it is convenient to avert fire accidents by one of the following ways: (i) admit/allow more fresh oxygen into the containment, which shall reduce the concentration of the fuel mixture. This shall subsequently change the LFL and UFL of the mixture and the flammable region will be converged sharply, (ii) alter the working temperature in the containment as temperature influences LFL and UFL limits of the fuel mixture. Hence, flammability diagrams give firsthand information for the safety personnel to understand whether the given fuel mixture lies within the flammable region or not. As an example, the flammability diagram of methane is drawn below.

Data given for methane (CH_4):
Flammability limit in air:
LFL = 5.3% of fuel in air; UFL = 15% of fuel in air
Flammability limit in pure oxygen:

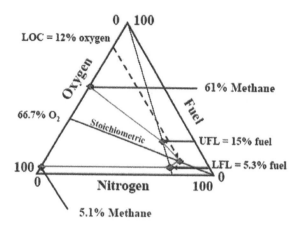

FIGURE 1.9   Flammability diagram of methane.

LFL = 5.1% of fuel in oxygen; UFL = 61% of fuel in oxygen
LOC = 12% oxygen
Stoichiometric point is at 66.7% of oxygen. Following the above steps, flammability diagram is drawn as shown in Figure 1.9.

1.10   DAMAGE CONSEQUENCES OF FIRE AND EXPLOSION

One of the most common methods used to quantify the consequences that arise from explosions is in terms of TNT (Chandrasekaran, 2016). TNT is an explosive with a few special characteristics. It can rapidly change its state from solid to hot-expanding gas state (Aven and Vinnem, 2007). Elements of carbon, oxygen, and nitrogen present in TNT produce highly stable substances with a strong bonding between them while it burns. However, as TNT explosions are chemically unstable, significant energy is not required to break this bond. In explosion damage scenario, it is a common practice to estimate the scaled distance of the spread of explosion energy. To compute this, equivalent mass of TNT needs to be estimated. Following are the steps involved to estimate the scaled distance:

Step 1: Estimate the flammable inventory present in the scenario. This is done by calculating the total mass (m) of the flammable fuel participating in the explosion.

Step 2: Determine the energy of explosion ΔH_c.

Step 3: Fix the efficiency of explosion process (η), which depends on the site-dependent parameters that initiate the explosion process. In the absence of any detailed studies, this factor can be assumed in the range (1%–15%).

Step 4: Compute the equivalent mass of TNT using the following relationship:

$$m_{TNT} = \frac{\eta \; m \; \Delta H_c}{E_{TNT}} \qquad (1.48)$$

where E_{TNT} is the energy of explosion of TNT (= 4,686 kJ/kg).

Step 5: Determine the scaled distance (Z_e) using the following relationship:

$$Z_e = \frac{r}{\sqrt[3]{m_{TNT}}} \tag{1.49}$$

Z_e is expressed in m/kg³, where r is the distance of the point of concern from the explosion source (in m).

Step 6: Based on the scaled distance estimated in the previous step, one can estimate the resulting overpressure (P_0), either graphically (see Figure 1.4) or using the following relationship:

$$P_0 = P_s \times P_a \tag{1.50}$$

where P_s is the scaled overpressure, which can be calculated using Figure 1.4 for the corresponding scaled distance. P_a is the atmospheric pressure, whose equivalent values for a variety of explosion activities are given in Table 1.8 (Chandrasekaran, 2016).

Alternatively, overpressure can also be computed using the following relationship:

$$P_0 = P_a \left\{ \frac{1,616\left[1+\left(\dfrac{Z_e}{4.5}\right)^2\right]}{\sqrt{\left[1+\left(\dfrac{Z_e}{0.048}\right)^2\right]} \times \left[1+\left(\dfrac{Z_e}{0.32}\right)^2\right] \times \left[1+\left(\dfrac{Z_e}{1.35}\right)^2\right]} \right\} \tag{1.51}$$

TABLE 1.8
Equivalent Atmospheric Pressure for Explosion Activities

Pressure P_a (kPa)	Quantifiable Damage Arising from Explosion
0.28	Loud noise, like glass breaking (about 143 dB)
0.69	Breakage of small windows
3.4–3.6	Windows shatter; occasional damage to window frames
4.8	Minor damage to structure
6.9–13.8	Significant damage to wooden furniture and asbestos sheets
15.8	Lower limit of significant damage
17.2	Complete destruction of brick masonry buildings
27.6	Rupture of claddings in industrial structures, oil tank ruptures, 50% probability of human fatality
34.5–48.2	Partial destruction of framed structures
62	Damage of loaded wagons of cargo trains
68.9	Total destruction of RC framed structures and damage to heavy machinery
75	Probability of 90% of human fatality; complete damage of steel structures

2 Material Properties

SUMMARY

Material properties are crucial for investigating fire-resistant capacity and enabling an appropriate design technique. Although many modern structural systems are designed to develop their capacity from geometric form, material strength is inherent in achieving the desired capacity. In this chapter, the mechanical properties of materials, namely steel, concrete, and functionally graded materials are discussed in detail. Material properties at elevated temperatures and various parameters that influence the strength at high temperatures are presented to enhance the understanding of the material behavior at high temperatures.

2.1 INTRODUCTION

In strategic structures like offshore platforms, nuclear power plants, and heavy industrial plants, there is a high probability of fire accidents due to the process complexities or stock of flammable inventories. Furthermore, under advanced design methods that include fire loads as a part of accidental loads, material strength plays a crucial role in guaranteeing the no-collapse mode of the structural system. For example, marine risers or pipelines that carry flammable chemicals at high temperatures and pressure undergo material degradation at elevated temperatures causing a silent threat to their serviceability. Though steel is one of the most favorable materials for heavy industrial structures and offshore platforms, alternate materials also show the equivalent performance of steel at elevated temperatures (Chandrasekaran et al., 2021).

Form-dominant structures like large domes, shell roofs, folded plates, cable-stayed bridges, floating offshore platforms, semi-submersibles, and nuclear power plants are unique and expensive in design, installation, and functions. In addition to the complexities that arise from the innovative geometry, the combination of several environmental loads, namely wave, wind, current, and ice make the design more complex. The legislation calls for periodic certification of their functional safety, which is by and large depends on the fact of material strength. While design procedures ensure no immediate collapse and catastrophic failure of structures even under critical loads, material degradation under the environment's influence poses a consistent challenge in assessing their reliability. See, for example, offshore structures; waves and currents are underestimated in a few design cases while fatigue and corrosion are still debatable subjects. All these factors result in a heavy penalty on the existing weight, which otherwise governs the design of form-dominant structures. Furthermore, using the input data of materials at elevated temperatures, fire is also one of the critical design considerations to achieve a high level of reliability.

DOI: 10.1201/9781003328711-2

Different types of materials used in building construction serve a variety of purposes, namely installation, repair, rehabilitation, corrosion, and protection. The choice of material from a wide range like composites, concrete, mild steel, copper, aluminum, fiber-reinforced plastics, titanium, and buoyancy materials depends on the purpose of construction; all of the above cannot be used for the same purpose (Chandrasekaran and Jain, 2016). Therefore, the selection of materials for building construction is a complex task as no single material characteristic shall help rank the materials. Detailed knowledge about their structural characteristics and relevant code compliance is necessary to arrive at a practical design. The behavior of materials under high temperature, fire, explosion, impact, and ice loads is another critical design consideration as structures are designed for a high level of reliability (Chandrasekaran and Gaurav, 2017; Chandrasekaran, 2015, 2016, 2017). Materials used in form-dominant structural systems should possess unique characteristics, besides, being compatible with extreme weather conditions. In addition, materials used in marine construction need to possess unique characteristics besides being compatible with extreme weather conditions. Therefore, design procedures are also dominated by material characteristics in addition to the critical combination of various environmental loads (Chandrasekaran and Bhattacharya, 2012; Chandrasekaran, 2014).

A strong bond exists between the engineering materials and the type of structures constructed using them. Referring to the guidelines suggested by various international and Indian codes of practice for structural design, material selection is only desirable but not mandatory. Under the given choice of a wide variety of materials available for use, a few parameters govern their choice: type of applications, properties under regular, cyclic loads, high and very low temperature, recycling characteristics, non-toxic nature and bio-friendly characteristics, sustainability for the service life, and the code compliance. In particular, under elevated temperatures, these complexities increase manifold. It is important to note that conventional design procedures are governed by the codes, but the use of advanced materials is always welcome. Therefore, the choice of appropriate material for construction becomes a critical engineering decision. The use of advanced materials is always welcome even if their strength-related characteristics are referent values in a particular model.

As discussed above, form-dominant design concepts demand special material characteristics. For example, large global displacements based on the design convention demand high ductility in the materials. Cyclic and other loads of reversal in nature demand high fatigue strength. As it is well understood that under fatigue loads, the magnitude of the stress only follows the severity of failure induced by the number of stress cycle reversals; even low-magnitude stress under a large number of cycles can cause failure that can be lesser than the yield strength. The degradation of material properties in the marine environment is a major concern. Some of them are chemical and strength degradation, loss of fatigue strength, poor performance due to high-stress concentrations, corrosion, and biofouling effects. Therefore, materials should possess survivability under accidental loads and the capacity to withstand large hydrostatic pressure, generated under hurricanes. An intelligent choice of material under such special loads as fire should focus on the fact that the material should not initiate a catastrophic failure. Even if the failure occurs at a lesser probability, it should not be sudden or immediate.

2.2   PROPERTIES OF STEEL

The properties of structural steel result from its chemical composition and method of manufacturing. The significant properties of steel from a designer's perspective are strength, toughness, ductility, durability, and weldability. These mechanical properties vary based on their chemical composition, heat treatment, and manufacturing processes. The addition of small quantities of other elements to its major constituent, iron, influences the mechanical properties of steel. While the strength can be increased by the addition of alloys such as manganese, niobium, and vanadium, ductility, toughness, and weldability will be compromised. Minimizing sulfur enhances ductility, while the addition of nickel improves toughness. It is therefore important to arrive at a carefully balanced chemical composition to ensure that desired structural characteristics are achieved. Similar to that, the alloying elements, heat treatment, and manufacturing processes are of critical importance to the performance of the steel. Rolled or formed steel shows higher strength, but increased thickness reduces its yield strength.

The effect of heat treatment, deployed in steel manufacturing, also governs its strength and toughness. The common manufacturing processes are as-rolled, normalized, normalized-rolled, thermo-mechanically rolled (TMR), and quenched and tempered (Q&T) steel. Steel cools as it is rolled, with a typical rolling finish temperature of about 750°C. If it is allowed to cool naturally, then it is called as-rolled steel. Normalized steel is manufactured by heating as-rolled steel up to about 900°C, held at that temperature for a specific time, before being allowed to cool naturally. Normalizing refines the grain size and enhances the toughness. Normalized-rolled steel is manufactured by rolling the steel at a temperature above 900°C, which eliminates the extra process of reheating the material. Normalized and normalized-rolled steels have an 'N' designation.

Thermo-mechanically rolled steel is manufactured at a lower rolling finish temperature of about 700°C. A greater force is required to roll the steel at lower temperatures, but the properties remain unaltered. This classification of steel has an 'M' designation. Q&T steel is manufactured at about 900°C at which it is rapidly cooled or 'quenched' to produce steel with high strength and hardness, but low toughness. This classification has a 'Q' designation. Quenching involves cooling a product rapidly by immersion directly into water or oil. Tempering is the second stage of heat treatment to temperatures below the austenitizing range, which softens the steel and also makes them tougher and more ductile. The use of high tensile steel can reduce the volume of steel, but the steel needs to be tough at operating temperatures. Therefore, higher-strength steels require improved toughness and ductility, which can be achieved only with low-carbon clean steels under the thermo-mechanical rolling process. Table 2.1 shows the minimum yield and tensile strength of commonly used hot-rolled steel grades in structural design (BS EN 10025-2, 2019); S designation, as seen in the table, indicates a carbon steel. The minimum yield strength and the tensile strength are taken as characteristic yield and characteristic ultimate strengths for design purposes (BS EN 1993-1-1, 2005).

TABLE 2.1

Minimum Yield and Tensile Strength of Hot-Rolled Steel

Grade	Yield Strength (N/mm²) for Nominal Thickness (in mm)				Tensile Strength (N/mm²) for Nominal Thickness (in mm)	
	$t \leq 16$	$16 \leq t \leq 40$	$40 \leq t \leq 63$	$63 \leq t \leq 80$	$3 \leq t \leq 100$	$100 \leq t \leq 150$
S275	275	265	255	245	410	400
S355	355	345	335	325	470	450

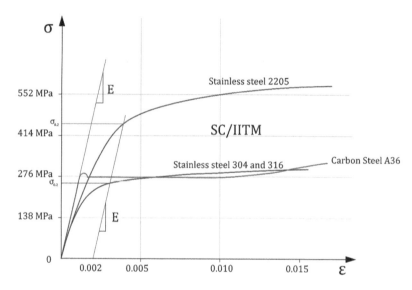

FIGURE 2.1   Stress-strain curve for carbon and stainless steel.

Stainless steel used for structural applications is classified into four groups, namely austenitic, ferritic, duplex, and martensitic. The former is the most commonly used stainless steel for construction. For example, Alloy 310 is austenitic stainless steel used for high-temperature and corrosion-resistant applications. The strength of stainless steel ranges from 170 to 450 N/mm². The austenitic type has yield strength lesser than carbon steel, while the duplex type has more. However, the strength ratio, which is the ratio of ultimate to yield strength, is higher in carbon steel, compared to stainless steel. Figure 2.1 shows the stress-strain curve for carbon steel and stainless steel. The variations in the ductility and strength can be seen in the figure. Table 2.2 shows the mechanical properties of various types of stainless steel used in structural engineering applications. Ductility indicates the strain or elongation capability under tensile loading. Ductility is desirable for various reasons, namely redistribution of stress at the ultimate limit state; reduced risk of fatigue crack propagation; and the fabrication processes of welding, bending, and straightening

TABLE 2.2
Mechanical Properties of Stainless Steel (EN 10088-4, 2009)

Description	Minimum 0.2% Proof Strength (N/mm²)	Ultimate Tensile Strength (N/mm²)	Percentage Elongation at Fracture (%)
Chromium nickel austenitic steel	200–210	520–570	45
Molybdenum chromium nickel austenitic steel	220	520–670	45
Duplex steel	450–460	640–850	25–30

2.2.1 CARBON STEEL AND ALLOYING ELEMENTS

There are three types of carbon steel, namely low-carbon, medium-carbon, and high-carbon steel; each type varies significantly in its properties. The designation of carbon steel is indexed as 10xx, which has xx% of carbon content. For example, 1,006 steel has a carbon content of 0.06%. Carbon content higher than 0.3% decreases its weldability below the desired value. Low-carbon steel has carbon content in the range of 0.05–0.3%. It has high weldability and workability, but low hardenability, machinability, strength-to-weight ratio, and wear resistance. Medium-carbon steel has carbon content in the range of 0.3–0.6%. It has medium to low weldability, high to medium surface hardening, strength-to-weight ratio, and machinability. They have specific applications to crankshafts and couplings. High-carbon steel has carbon content in the range of 0.6–1%. It possesses very low weldability, low machinability, and workability while having a high hardenability, strength-to-weight ratio, and wear resistance. Common applications are springs, alloying elements, and cutting tools.

A few alloying elements of steel influence its mechanical properties, significantly. Aluminum improves case-hardening through nitriding, bismuth improves machinability, boron improves hardenability, chromium improves hardenability and corrosion resistance, copper improves corrosion resistance, lead improves machinability, manganese reduces brittleness and removes excess oxygen and increases hardenability, molybdenum improves toughness, nickel improves toughness and corrosion resistance, silicon improves strength, sulfur improves machining, titanium reduces martensitic hardness in chromium steel, tungsten increases melting point, and vanadium improves strength and ductility and also increases toughness at high temperatures.

2.3 PROPERTIES AT HIGH TEMPERATURE

Steel is one of the most favorite materials for the construction of form-dominant structures, in particular. But its behavior at high temperatures makes the design different from those of the conventional ones. Young's modulus, proportional limit, and yield strength decrease with the increase in temperature (Chandrasekaran, 2019, 2020a, 2020b; Chandrasekaran et al., 2021). These variations as discussed in Eurocode are shown in Figures 2.2–2.4, respectively. The corresponding MATLAB codes used to

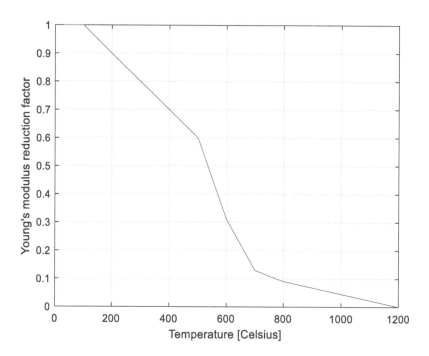

FIGURE 2.2   Young's modulus reduction in steel.

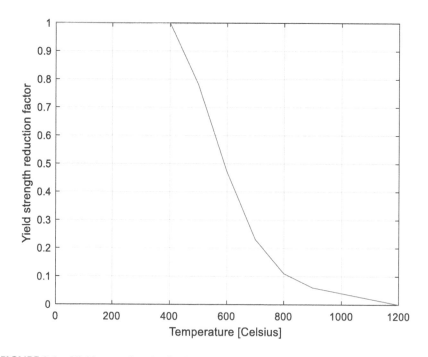

FIGURE 2.3   Yield strength reduction in steel.

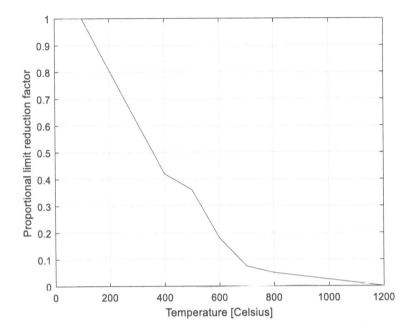

FIGURE 2.4　Proportional limit reduction in steel.

plot the figures are also given below. It can be observed that steel loses about half of its strength at around (550°C), which is referred to as the critical temperature of steel. It is to be noted that other codes (e.g., ASCE) do not consider a separate model for the proportional limit, and hence are not able to capture viscoelastic behavior near the transition from elastic to inelastic zones in steels at high temperature.

```
steel_modulus_reduction.m
function kE = steel_modulus_reduction(T)
% Computes temperature-dependent reduction in Young's modulus
of steel.
% Uses Eurocode 3 definition
% T: vector of temperatures where conductivity is needed
(Celsius)
% kE: vector of reduction factors for Young's modulus
(dimensionless)
n = length(T);
kE = zeros(n, 1);
 % Eurocode 3 Table:
Tpoints = [20; [100:100:1200]'];
kEpoints = [1; 1; .9; .8; .7; .6; .31; .13; .09; .0675; .045;
.0225; 0];
 kE = interp1(Tpoints, kEpoints, T);


steel_yield_strength_reduction.m
function ky = steel_yield_strength_reduction(T)
```

```
% Computes temperature-dependent reduction in yield strength
of steel.
% Uses Eurocode 3 definition
% T: vector of temperatures where conductivity is needed
(Celsius)
% ky: vector of reduction factors for Young's modulus
(dimensionless)
n = length(T);
ky = zeros(n, 1);
 % Eurocode 3 Table:
Tpoints = [20; [100:100:1200]'];
kypoints = [1; 1; 1; 1; 1; .78; .47; .23; .11; .06; .04; .02;
0];
ky = interp1(Tpoints, kypoints, T);
```

steel_prop_limit_reduction.m
```
function kp = steel_prop_limit_reduction(T)
% Computes temperature-dependent reduction in proportional
limit of steel.
% Uses Eurocode 3 definition
% T: vector of temperatures where conductivity is needed
(Celsius)
% kp: vector of reduction factors for proportional limit
(dimensionless)
n = length(T);
kp = zeros(n, 1);
 % Eurocode 3 Table:
Tpoints = [20; [100:100:1200]'];
kppoints = [1; 1; .807; .613; .42; .36; .18; .075; .05; .0375;
.025; .0125; 0];
kp = interp1(Tpoints, kppoints, T);
```

The other properties of carbon steel under high temperatures also vary significantly and need to be considered in the design. Most of the thermal properties of structural steel are similar to that of the reinforcing steel as well. The specific heat of steel is 450 J/kg K at room temperature, which increases up on heating up to a temperature of around 730°C. Around this temperature, a phase change takes place in which the lattice structure of steel changes from being face-centered to body-centered. The variation of specific heat of steel as per Eurocode is shown in Figure 2.5. A MATLAB function that was used to generate this figure is included below the figure. It can be used in computerized calculations to obtain the specific heat of steel when required.

steel_specific_heat.m
```
function c = steel_specific_heat(T)
% Computes temperature-dependent specific heat of steel.
% Uses Eurocode 3 definition
% T: vector of temperatures where conductivity is needed
(Celsius)
% c: vector of specific heat (J/kgK)
```

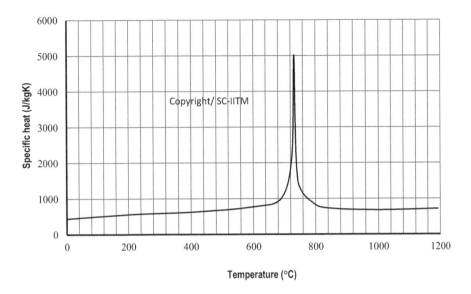

FIGURE 2.5 Variation of specific heat of carbon steel with temperature.

```
n = length(T);
c = ones(n, 1)*650;
for i = 1:n
    if T(i) < 600
        c(i) = 425 + 7.73e-1*T(i) - 1.69e-3*T(i)^2 +
2.22e-6*T(i)^3;
    elseif T(i) < 735
        c(i) = 666 + 13002/(738-T(i));
    elseif T(i) < 900
        c(i) = 545 + 17820/(T(i) - 731);
    end
end
```

Some studies (Idogaki, 1979) reported that the peak at 730°C is not so. However, that does not impact the design calculations significantly. Wielgosz et al. (2014) reported that steel shows another peak in specific heat in the range of 1450–1500°C. This also has no impact as long as structural design calculations are concerned.

The thermal conductivity of steel is around 50 W/mK at room temperature, which decreases linearly up on heating up to a temperature of around 800°C. Beyond this temperature, it remains constant (Chandrasekaran et al., 2021). Although steel is one of the most preferred construction materials, strength degradation and de-crystallization of steel at higher temperatures remain a significant challenge. The variation of thermal conductivity of steel as per Eurocode is shown in Figure 2.6. A MATLAB function that was used to generate this figure is included below the figure. It can be used in computerized calculations to obtain the thermal conductivity of steel when required. It is important to note that high-alloy steels have lower thermal conductivities than low-alloy steels (The, O. 1934).

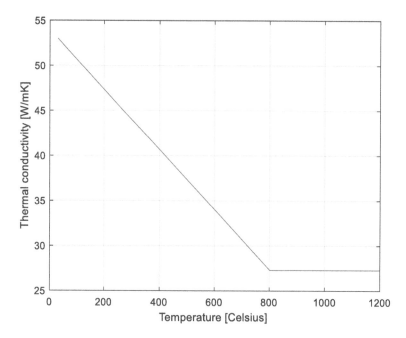

FIGURE 2.6 Variation of thermal conductivity of steel with temperature.

`steel_thermal_conductivity.m`
```
function k = steel_thermal_conductivity(T)
% Computes temperature-dependent thermal conductivity of
steel.
% Uses Eurocode 3 definition
% T: vector of temperatures where conductivity is needed
(Celsius)
% k: vector of thermal conductivities (W/mK)
n = length(T);
k = ones(n, 1)*27.3;
for i = 1:n
    if T(i) < 800
        k(i) = 54 - 3.33e-2 * T(i);
    end
end
```

Thermal strain variation as a function of time is shown in Figure 2.7. Thermal strain is defined as the ratio of the change in length due to temperature rise to its original length at ambient temperature. MATLAB program used to plot the variation is also given below the figure, which is useful in design calculations and numerical modeling of such variations.

`steel_thermal_strain.m`
```
function eT = steel_thermal_strain(T)
% Computes the thermal strain of steel.
% Uses Eurocode 3 definition
```

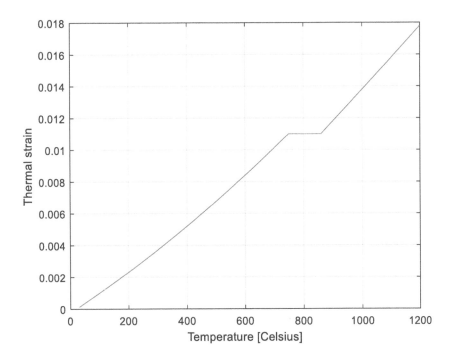

FIGURE 2.7 Variation of thermal strain of steel with temperature.

```
% T: vector of temperatures where conductivity is needed
(Celsius)
% eT: vector of thermal strain (dimensionless)
 n = length(T);
eT = zeros(n, 1);
for i = 1:n
    if T(i) < 750
        eT(i) = 1.2e-5*T(i) + 4e-9*T(i)^2 - 2.416e-4;
    elseif T(i) <= 860
        eT(i) = 1.1e-2;
    else
        eT(i) = 2e-5*T(i) - 6.2e-3;
    end
end
```

2.4 REINFORCING STEEL

2.4.1 MECHANICAL PROPERTIES

Eurocode specifies a robust temperature-dependent constitutive behavior for reinforcing steel. The relations are not reproduced directly. It is parametrized by six parameters, as shown in Figure 2.8. A MATLAB program to generate the stress-strain relation is provided in rebar_constitutive.m. These relations at different temperatures are plotted in Figure 2.9.

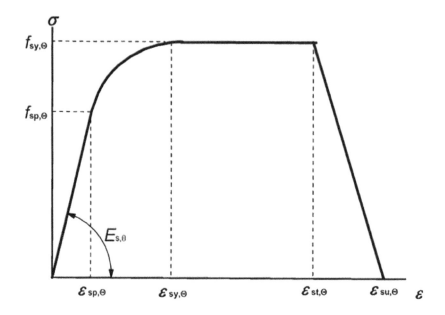

FIGURE 2.8   Constitutive behavior of reinforcing steel.

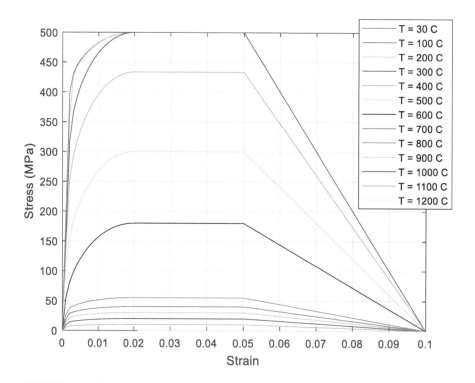

FIGURE 2.9   Constitutive behavior of reinforcing steel at different temperatures.

TABLE 2.3
Constitutive Behavior of Rebar

Range	Stress $\sigma(\theta)$	Tangent Modulus
$\varepsilon_{sp,\theta}$	$\varepsilon\, E_{s,\theta}$	$E_{s,\theta}$
$\varepsilon_{sp,\theta} \leq \varepsilon \leq \varepsilon_{sy,\theta}$	$f_{sp,\theta} - c + (b/a)[a^2 - (\varepsilon_{sy,\theta} - \varepsilon)^2]^{0.5}$	$\dfrac{b(\varepsilon_{sy,\theta} - \varepsilon)}{a\left[a^2 - (\varepsilon - \varepsilon_{sy,\theta})^2\right]^{0.5}}$
$\varepsilon_{sy,\theta} \leq \varepsilon \leq \varepsilon_{st,\theta}$	$f_{sy,\theta}$	0
$\varepsilon_{st,\theta} \leq \varepsilon \leq \varepsilon_{su,\theta}$	$f_{sy,\theta}[1 - (\varepsilon - \varepsilon_{st,\theta})/(\varepsilon_{su,\theta} - \varepsilon_{st,\theta})]$	-
$\varepsilon = \varepsilon_{su,\theta}$	0,00	-

Parameter[*)]	$\varepsilon_{sp,\theta} = f_{sp,\theta}/E_{s,\theta}$	$\varepsilon_{sy,\theta} = 0,02$	$\varepsilon_{st,\theta} = 0,15$	$\varepsilon_{su,\theta} = 0,20$
	Class A reinforcement :		$\varepsilon_{st,\theta} = 0,05$	$\varepsilon_{su,\theta} = 0,10$

Functions	$a^2 = (\varepsilon_{sy,\theta} - \varepsilon_{sp,\theta})(\varepsilon_{sy,\theta} - \varepsilon_{sp,\theta} + c/E_{s,\theta})$
	$b^2 = c(\varepsilon_{sy,\theta} - \varepsilon_{sp,\theta})E_{s,\theta} + c^2$
	$c = \dfrac{\left(f_{sy,\theta} - f_{sp,\theta}\right)^2}{(\varepsilon_{sy,\theta} - \varepsilon_{sp,\theta})E_{s,\theta} - 2(f_{sy,\theta} - f_{sp,\theta})}$

The constitutive relationship for rebar properties is given in Table 2.3.

Matlab program to plot the constitutive behavior of rebar
rebar_constitutive.m

```
function [stress, strain] = rebar_constitutive(T)
% Uses Eurocode 3 definition
% T: temperature for which constitutive relation is needed
(Celsius)
 % Cold rebar constitutive parameters
fp_20 = 400;    %proportional limit, MPa
fy_20 = 500;    %yield strength, MPa
E_20  = 200e3;  %Young's modulus, MPa
% Calculate reduction factors for current temperature
[k_fy, k_fp] = rebar_strength_reduction(T);
kE = steel_modulus_reduction(T);
% Parameters of the constitutive relation curve (from the code)
fsp_T = k_fp * fp_20;
fsy_T = k_fy * fy_20;
E_s_T = kE * E_20;
 e_sp_T = fsp_T / E_s_T;
e_sy_T = 0.02;
e_st_T = 0.05;
e_su_T = 0.10;
 % Don't change beyond this line
c = (fsy_T - fsp_T)^2 / ((e_sy_T - e_sp_T)*E_s_T - 2*(fsy_T
- fsp_T));
```

```
b = sqrt( c*(e_sy_T - e_sp_T)*E_s_T + c^2 );
a = sqrt((e_sy_T - e_sp_T)*(e_sy_T - e_sp_T + c/E_s_T));
 strain = 0:1e-3:e_su_T;
n = length(strain);
stress = zeros(n, 1);
for i = 1:n
   e = strain(i);
   if e < e_sp_T
      stress(i) = e * E_s_T;
   elseif e <= e_sy_T
      stress(i) = fsp_T - c + (b/a)*(a^2 - (e_sy_T - e)^2)^0.5;
   elseif e <= e_st_T
      stress(i) = fsy_T;
   elseif e < e_su_T
      stress(i) = fsy_T * (1 - (e - e_st_T)/(e_su_T - e_st_T));
   else
      stress(i) = 0;
   end
end
```

Eurocode provides different values of reduction factor for class N when no experimental data are available, and class X when experimental data are available. Figure 2.10 shows the strength reduction factors for reinforcing and pre-stressing steel at high temperature. MATLAB program used to plot the figure is also given below.

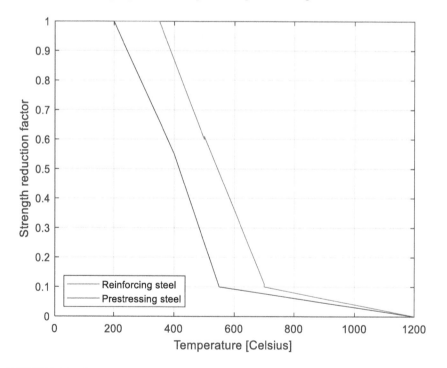

FIGURE 2.10   Strength reduction factor of reinforcing and pre-stressing steel.

rebar_strength_reduction.m

```
function [ks, kp] = rebar_strength_reduction(T)
% Computes temperature-dependent reduction in strength of
rebar steel.
% Uses Eurocode 3 definition
% T: vector of temperatures where conductivity is needed
(Celsius)
% ks: vector of reduction factors for reinforcing steel
strength (dimensionless)
% kp: vector of reduction factors for prestressing steel
strength (dimensionless)
n = length(T);
ks = zeros(n, 1); kp = ks;
% reinforcing steel
for i = 1:n
    if T(i) <= 350
        ks(i) = 1;
    elseif T(i) <= 500
        ks(i) = 1 - 0.4*(T(i) - 350)/150;
    elseif T(i) <= 700
        ks(i) = 0.61 - 0.5*(T(i) - 500)/200;
    else
        ks(i) = 0.1 - 0.1*(T(i) - 700)/500;
    end
end
% prestressing steel
for i = 1:n
    if T(i) <= 200
        kp(i) = 1;
    elseif T(i) <= 400
        kp(i) = 1 - 0.45*(T(i) - 200)/200;
    elseif T(i) <= 550
        kp(i) = 0.55 - 0.45*(T(i) - 400)/150;
    else
        kp(i) = 0.1 - 0.1*(T(i) - 550)/650;
    end
end
```

2.5 CONCRETE

2.5.1 MECHANICAL PROPERTIES

Concrete behaves differently under compression and tension. The parametrized constitutive behavior of concrete under compression is shown in Figure 2.11. It has three parameters as mentioned below. The post-peak portion can be idealized as linear or non-linear curve depending on the available data. Eurocode's constitutive relation is relatively straightforward and practical when compared with the relations given by other researchers (Chang et al., 2006; Kodur, 2014). Eurocode enables both linear and non-linear post-peak behavior, while ASCE suggests non-linear behavior for the post-peak portion (Kodur, 2014). As per Eurocode, concrete tensile strength reduces

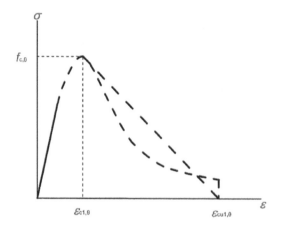

Range	Stress $\sigma(\theta)$
$\varepsilon \leq \varepsilon_{c1,\theta}$	$\dfrac{3\varepsilon f_{c,\theta}}{\varepsilon_{c1,\theta}\left[2+\left(\dfrac{\varepsilon}{\varepsilon_{c1,\theta}}\right)^3\right]}$
$\varepsilon_{c1(\theta)} < \varepsilon \leq \varepsilon_{cu1,\theta}$	For numerical purposes a descending branch should be adopted. Linear or non-linear models are permitted.

FIGURE 2.11 Constitutive behavior of concrete under compression.

to zero at 600°C, although the experiments conducted by Chang et al. (2006) suggest that the tensile strength of concrete is reduced to 35–40% at this temperature. As concrete is generally weak in tension, the tensile strength is neglected for design-related calculations. Hence, such an assumption does not have any impact on the design calculations through code-based approaches. Furthermore, detailed constitutive behavior is generally not considered in usual design considerations; instead, only the tensile strength is utilized where required.

The compressive and tensile strengths of concrete decrease with the rise in temperature. While the Eurocode suggests different compressive strength reduction factors for siliceous (granite, sandstone) and carbonate aggregates (dolomite, limestone), American Society for Testing and Materials (ASTM) prescribes a single reduction factor. For tensile strength, both the codes provide the use of a common reduction factor regardless of the kind of aggregate used. MATLAB codes for computation of reduction factors for compressive and tensile strengths of concrete are shown in concrete _ compression _ reduction.m and concrete _ tensile _ reduction.m, respectively. A graphical representation of these reduction factors is shown in Figure 2.12.

```
concrete_compression_reduction.m
function [kcs, kcc] = concrete_compression_reduction(T)
% Uses Eurocode 3 definition
```

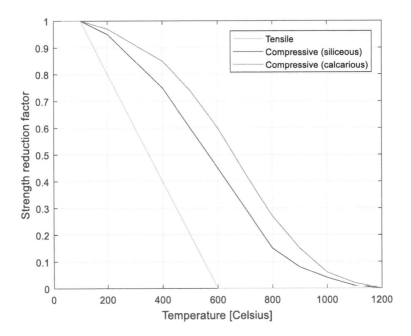

FIGURE 2.12 Reduction factors for compressive and tensile strength of concrete at high temperature.

```
% T: vector of temperatures where conductivity is needed
(Celsius)
% kcs: vector of compression strength reduction factor
(dimensionless)
%        (siliceous aggregate)
% kcc: vector of compression strength reduction factor
(dimensionless)
%        (calcareous aggregate)
 n = length(T);
% Eurocode 3 Table:
Tpoints = [20; [100:100:1200]'];
kcspoints = [1; 1; .95; .85; .75; .6; .45; .3; .15; .08; .04;
.01; 0];
kccpoints = [1; 1; .97; .91; .85; .74; .6; .43; .27; .15; .06;
.02; 0];
kcs = interp1(Tpoints, kcspoints, T);
kcc = interp1(Tpoints, kccpoints, T);
```

concrete_tensile_reduction.m
```
function kt = concrete_tensile_reduction(T)
% Uses Eurocode 3 definition
% T: vector of temperatures where conductivity is needed
(Celsius)
% kt: vector of tensile strength reduction factor
(dimensionless)
```

```
  n = length(T);
kt = zeros(n, 1);
  for i = 1:n
    if T(i) <= 100
        kt(i) = 1;
    elseif T(i) <= 600
        kt(i) = 1 - (T(i) - 100)/500;
  end
end
```

It can be noted that concrete loses about half of its compressive strength at 600°C. Eurocode model is generally conservative when compared to the experimental values. Further, the reduction of compressive strength given by Eurocode is not linear with respect to rise in temperature, unlike the ASTM model. The compressive strength of concrete reduces from 100°C as per the Eurocode, while ASCE suggests no reduction in the compressive strength up to 400°C (Kodur, 2014). ASCE also suggests that the compressive strength of concrete reduces to zero at about 900°C. Thus, the temperature rate of strength reduction as per the ASCE model is more rapid as compared to that of the Eurocode model.

2.5.2 THERMAL PROPERTIES

The specific heat of concrete depends on its moisture content within the temperature range when water exists in vapor phase. The evaporation begins at about 100°C, resulting in a sharp rise in the specific heat. It gradually reduces as the temperature reaches 200°C. The specific heat model as per Eurocode can be calculated for different moisture content and temperature conditions for concrete, as given in the MATLAB program concrete _ specific _ heat.m. A graphical representation of the specific heat of concrete for different moisture contents is shown in Figure 2.13.

```
concrete_specific_heat.m
function c = concrete_specific_heat(T, u)
% Uses Eurocode 3 definitionconcr
% T: vector of temperatures where conductivity is needed
(Celsius)
% u: moisture content of concrete
% c: vector of specific heat (J/kgK)
 n = length(T);
c = zeros(n, 1);
% Table for cpeak interpolation
cp_peak_moisture_points = [0; 1.5; 3];
cp_peak_values = [900; 1470; 2020];
cpeak = interp1(cp_peak_moisture_points, cp_peak_values, u);
  for i = 1:n
    if T(i) <= 100
        c(i) = 900;
    elseif T(i) <= 200
        % values depend on moisture content in this temperature
        range
```

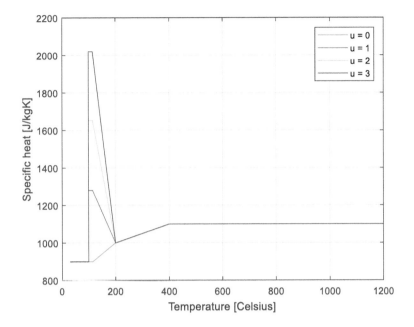

FIGURE 2.13 Specific heat of concrete.

```
    if T(i) <= 115
        c(i) = cpeak;
    else
        c(i) = interp1([115; 200], [cpeak; 1000], T(i));
    end
  elseif T(i) <= 400
    c(i) = 1000 + (T(i) - 200)/2;
  else
    c(i) = 1100;
  end
end
```

While ASCE provides different models for assessing the thermal capacity (i.e., ρc) of concrete, Eurocode provides different models for assessing the specific heat (c) and density (ρ) separately. The model for reduction in density as per Eurocode is given in the MATLAB program concrete density _ reduction.m and shown in Figure 2.14.

concrete_density_reduction.m
```
function p = concrete_density_reduction(T)
% Uses Eurocode 3 definition
% T: vector of temperatures where conductivity is needed
(Celsius)
% p: vector of density reduction factor (dimensionless)
  n = length(T);
p = zeros(n, 1);
```

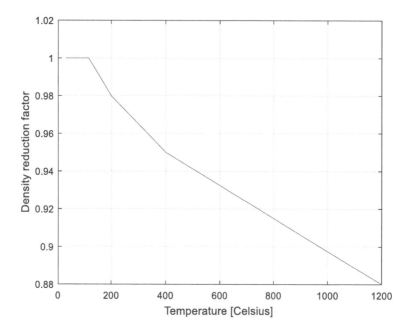

FIGURE 2.14   Density reduction of concrete.

```
for i = 1:n
   if T(i) <= 115
      p(i) = 1;
   elseif T(i) <= 200
      p(i) = 1 - 0.02*(T(i) - 115)/85;
   elseif T(i) <= 400
      p(i) = 0.98 - 0.03*(T(i) - 200)/200;
   else
      p(i) = 0.95 - 0.07*(T(i) - 400)/800;
   end
end
```

Thermal conductivity of concrete decreases gradually with the rise in temperature. Eurocode suggests relations for upper and lower limits of the thermal conductivity of concrete; the model is given in concrete _ thermal _ conductivity.m and shown in Figure 2.15. Factors affecting the thermal conductivity of concrete (in order of effectiveness) include moisture content, recycled coarse aggregate replacement ratio, temperature, volume fractions of aggregates, types of aggregate, and types of admixtures (Pan et al., 2017).

concrete_thermal_conductivity.m
```
function [kUpper, kLower] = concrete_thermal_conductivity(T)
% Uses Eurocode 3 definition
% T: vector of temperatures where conductivity is needed (Celsius)
% kUpper: vector of upper limit of thermal conductivities (W/mK)
```

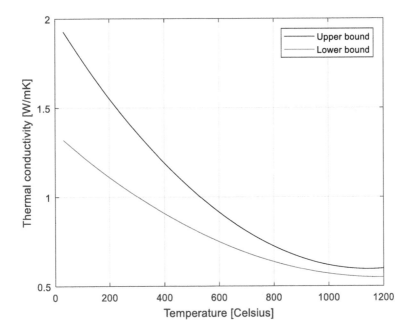

FIGURE 2.15 Thermal conductivity of concrete.

```
% kLower: vector of lower limit of thermal conductivities (W/mK)
 kUpper = 2 - 0.2451*(T/100) + 0.0107*(T/100).^2;
kLower = 1.36 - 0.136*(T/100) + 0.0057*(T/100).^2;
```

Thermal strain of concrete also depends on the type of aggregates used. Eurocode model is given in concrete _ thermal _ strain.m, and the variation is shown in Figure 2.16.

concrete_thermal_strain.m
```
function [es, ec] = concrete_thermal_strain(T)
% Uses Eurocode 3 definition
% T: vector of temperatures where conductivity is needed
(Celsius)
% es: vector of thermal strain for concrete with siliceous
aggregate (dimensionless)
% ec. vector of thermal strain for concrete with calcareous
aggregate (dimensionless)
 n = length(T);
es = zeros(n, 1); ec = es;
 for i = 1:n
    if T(i) <= 700
        es(i) = -1.8e-4 + 9e-6 * T(i) + 2.3e-11 * T(i)^3;
    else
        es(i) = 14e-3;
    end
```

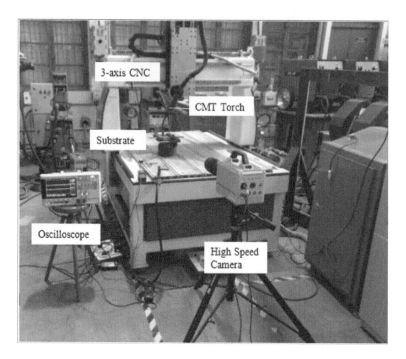

FIGURE 2.16   WAAM unit used to manufacture FGM. (Chandrasekaran et al., 2022.)

```
end
for i = 1:n
   if T(i) <= 805
      ec(i) = -1.2e-4 + 6e-6 * T(i) + 1.4e-11 * T(i)^3;
   else
      ec(i) = 12e-3;
   end
end
```

2.6   FUNCTIONALLY GRADED MATERIALS

When structural members are subjected to a wired combination of environmental loads in the presence of high-temperature and high-pressure conditions, conventional materials including steel fail to satisfy all the durability and strength requirements. In such cases, different materials good at certain mechanical properties are chosen and manufactured in a continuously graded manner. Functionally graded materials are novel materials, manufactured using special processes and not physically interconnected/welded to form a sandwich material. Functionally-Graded Materials (FGMs) are intended to overcome the damages posed by delamination as they do not possess any distinct interface between the constituent materials (Chandrasekaran, 2020a, 2020b).

FGMs are increasingly used in aerospace, defense, and medical applications for specific functional requirements. In the recent past, detailed studies carried out on

FGM prove their probable application in offshore platforms as well (Chandrasekaran, 2019, 2020a, 2020b). The creation of thermal barriers, anti-oxidation coatings, and cemented-carbide cutting tools is some of the successful engineering applications where FGMs are used. Based on the method of manufacturing, the constituent materials of FGM can vary continuously or in a stepwise manner. In both the manufacturing methods, there exists no distinct layer of interface between the constituent materials; the developed product will be a homogeneous one but possess different mechanical characteristics along its cross section.

Functional grading utilizes the salient advantages of the composition of materials. Metallic FGMs using nickel-based superalloys (Yeo et al., 1998), titanium (Gao and Wang, 2000), and stainless steel (Rajan et al., 2010) are successfully fabricated by several manufacturing techniques. FGM can be manufactured by many conventional methods, namely centrifugal casting (Shishkovsky et al., 2012), spark plasma sintering (Yuan et al., 2012; Jin et al., 2005), laser deposition (Ubeyli et al., 2014), and powder metallurgy (Chen et al., 2017; Bermingham et al., 2015). The advancements in material science and improved additive techniques enable the manufacturing of FGM using additive manufacturing techniques, namely electron beam melting, selective laser melting, direct-laser melting process, and wire arc additive manufacturing (WAAM).

FGM is a novel material, which is manufactured by functionally grading two-metal components, chosen based on specific functional requirements. Let us consider a marine riser as an example for discussion. The hydrocarbons flowing through marine risers pose a continuous threat of corrosion. At the same time, due to the high-pressure and high-temperature conditions, they are also subjected to high-stress intensities. Hence, strength and corrosion resistance shall be chosen as the intended functions for choosing the constituent materials to manufacture FGM. However, manufacturing FGM, by combining one or more materials of designers' choice, is a complex task as the manufacturing process imposes significant challenges in achieving the desired properties of FGM. As a common practice, materials of desired characteristics are chosen, and their geometric compositions (not the metallurgical composition) in terms of thickness and number of layers are varied continuously across the cross section to manufacture FGM. Thus, the composition and microstructure are altered along the cross section to generate the desired property gradient. It is intended to utilize completely the mechanical, metallurgical, and structural properties of the original materials while forming the FGM (Chandrasekaran and Hari, 2022; Chandrasekaran et al., 2019, 2020, 2022).

The WAAM enables the metallurgical composition of user-defined materials by a stepwise addition (Martina et al., 2012). The component metals are deposited in layers of wires, which are advanced using a secondary wire-feeder at the desired speed. A high-pulse current is supplied to form an arc between the electrode wires and the substrate, resulting in the melting of the filler tip of these advancing wires. A stainless-steel substrate is used to deposit the materials. Figure 2.16 shows various components of the WAAM unit, namely the Cold Metal Transfer (CMT) torch, the substrate, and the CNC machine integrated with the torch. The deposition parameters for the WAAM process are based on the constituent materials and the appropriate fillers (Chandrasekaran and Hari, 2022; Chandrasekaran et al., 2019, 2020).

2.6.1 FGM FOR HIGH TEMPERATURE

FGM is manufactured on a lab scale by WAAM. The materials chosen are duplex stainless steel and carbon-manganese steel, which possess strength and corrosion resistance properties, as specific functional requirements. Fronius TM CMT Transpulse Synergic 4,000 is used as a power source for the deposition process. A three-axis Computer Numeric Control (CNC) machine is integrated into the welding torch, as shown in Figure 2.16. The welding torch is programmed using G-codes in the desired *X, Y,* and *Z* coordinates to obtain the weaving pattern of deposition; a bidirectional weaving strategy is used for deposition. More details are discussed by the researchers on this specific application. See Chandrasekaran et al. (2022) and Chandrasekaran (2020). Using the deposition parameters, three layers of duplex stainless steel are deposited on the stainless-steel substrate, and five layers of carbon-manganese steel are deposited on the previously deposited duplex stainless steel forming an FGM as shown in Figure 2.17.

Tension tests are carried out on three samples of ASTM E8 specimens, extracted from the FGM build; the test specimen is shown in Figure 2.18. Figure 2.19 shows the direction of tensile load under the Universal Testing Machine (UTM). Please note that the direction of tensile load is aligned along with the interface of the material, though a distinct physical interface is not seen in the FGM sample, extracted from the build. Figure 2.20 shows the test setup to derive the mechanical properties of the FGM.

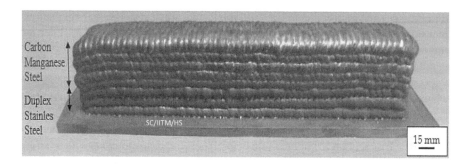

FIGURE 2.17   FGM comprising duplex stainless steel and carbon-manganese steel.

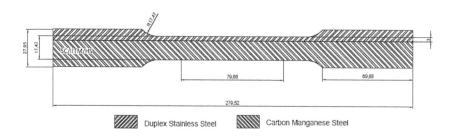

FIGURE 2.18   ASTM test specimen extracted from FGM build.

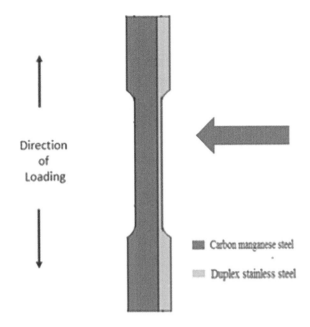

Direction
of
Loading

■ Carbon manganese steel

▦ Duplex stainless steel

FIGURE 2.19 Tensile load on the specimen in UTM.

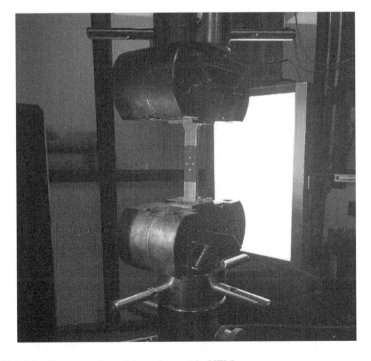

FIGURE 2.20 Test setup for axial tension test in UTM.

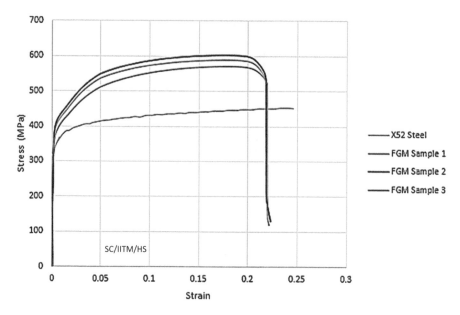

FIGURE 2.21   Stress-strain curves of FGM samples and X52 steel.

A gauge width of 17.47 is maintained in the sample, as shown in the figure, which consists of 3 mm of duplex stainless steel and 14.47 mm of carbon-manganese steel. A UTM Zwick Roell Z100 integrated into a video extensometer is used to measure the longitudinal tensile strain and the transverse compressive strains. The test specimen is examined under X-ray computed tomography and observed with no porosity and microcracks in all three orientations. It also confirms the successful application of WAAM for fabricating FGM to obtain a completely homogeneous and solid build. Fractography is carried out on the sample after the tensile test by a scanning electron microscope Quanta 200F using secondary electron imaging mode. The FGM material is also tested under elevated temperature as per ASTM E21. The test specimens are soaked for 30 min at 200°C before the test is carried out.

The tensile tests on three FGM samples are carried out at a strain rate of 1×10^{-3} per s. The stress-strain curves obtained under the longitudinal axis are shown in Figure 2.21 along with their comparison to X52 steel at room temperature. As seen from the figure, the ultimate strength measured is 602 MPa, while the yield strength is 403 MPa. The modulus of elasticity is computed as 213 GPa. Figure 2.22 shows the stress-strain curve obtained from the tensile test along the lateral direction. The strain obtained in the lateral direction is limited to the extensometer capacity to capture due to the white dots moving along the longitudinal direction. Table 2.4 shows the comparison of mechanical properties of FGM samples at the room temperature and the elevated temperature. Table 2.5 compares the mechanical properties of FGM with X52 steel. The test specimen after the high-temperature tensile test is shown in Figure 2.23, which shows no delamination between the functionally graded layers. As seen in the table, the yield strength and elongation at high

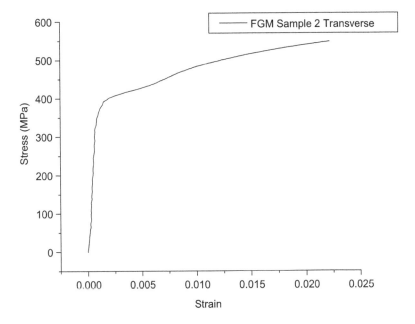

FIGURE 2.22   Stress-strain curve of FGM sample in transverse direction.

TABLE 2.4
Comparison of Mechanical Properties During High-Temperature Test

Material Parameters	FGM at Room Temperature	FGM at 200°C
Yield strength (MPa)	390.66 ± 12.23	339.93
Ultimate strength (MPa)	587.66 ± 12.76	468.19
% Elongation	22.31 ± 0.11	18.9

TABLE 2.5
Comparison of the Mechanical Properties

Material Parameters	X52 Steel	FGM
Young's modulus (GPa)	210	209.66 ± 4.48
Yield strength (MPa)	358	390.66 ± 12.23
Ultimate strength (MPa)	453	587.66 ± 12.76
Strength ratio	1.265	1.50 ± 0.02
Ductility ratio	32.207	45.47 ± 0.82
Tensile toughness (J/m³)	104.92	120.50 ± 2.84
Poisson ratio	0.3	0.30 ± 0.07
% Elongation	21	22.31 ± 0.11

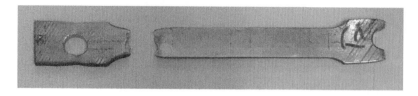

FIGURE 2.23 Fractured specimen after high-temperature tensile test.

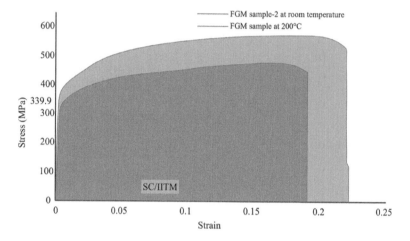

FIGURE 2.24 Toughness of FGM at normal and elevated temperatures.

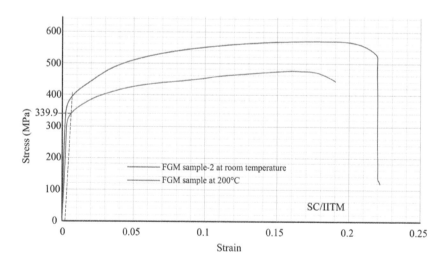

FIGURE 2.25 Stress-strain curves of FGM sample at normal and elevated temperatures.

temperature are close to that of X52 at room temperature while the ultimate strength is decreased. Figures 2.24 and 2.25 show the toughness and stress-strain curve of FGM at elevated temperatures. Yield stress computed at 0.2% proof stress can also be observed in the figure.

3 Fire Design

SUMMARY

This chapter discusses the key ideas behind the structural design for fire including presenting a view of fire as a hazard, characterization of fire as a load, temperature calculations, and structural design calculations. Some statistics related to damages caused by fire are also included to emphasize the need of considering fire as a load although it is an accidental occurrence. It is to be noted that the probability of occurrence of fire in a building is significantly greater than that of the building experiencing a major earthquake during its design life. Hence, appropriate treatment of fire scenarios in the design of structures should be mandatory. This chapter reviews the basic concepts of combustion of fuel in building fires and provides a brief idea about calorimetry used to determine the burning behavior of the fuel materials. It also discusses the concepts related to the quantification of the fire load. Fire development process within the compartment and various empirical fire development models that consider fire load and ventilation conditions are also discussed in detail. Additionally, the chapter also discusses temperature progression at different stages of fire in terms of standard and parametric fire time-temperature curves.

3.1 FIRE AS A HAZARD

Infrastructural facilities such as buildings, transportation systems, and offshore platforms are designed to serve a long service life during which they might get exposed to various extreme loads arising from landslides, earthquakes, floods, hurricanes, or fire. Therefore, these facilities must be designed in a manner so as to withstand the effects of such events. Fire can occur primarily because of accidents caused by cooking, electric short circuits (perhaps the most common cause of fires), smoking, use of faulty equipment, arson, lightning strike, or it can be a secondary effect of other hazards such as earthquakes or blasts. Since all of the aforementioned kinds of loads are accidental, their probability of occurrence is quite small; consequently, the probability of their combined occurrence (e.g., post-earthquake fire) is even smaller. Thus, consideration of fire as a load is not combined with any other accidental load (such as earthquake and hurricane).

Fire poses risks to human lives and property. In early stages of a building fire, life safety is important as all the egress and rescue operations should ideally be completed during this time. It is well established that majority of deaths due to fire do not occur due to burning but due to asphyxiation caused by the lack of oxygen experienced by people trapped in a compartment fire. Even small doses of carbon monoxide (as small as 15 ppm) can cause unconsciousness leading to death in case of prolonged exposure. The middle to later stages of a fire are crucial from the perspective of property loss. The sooner a fire can be controlled, the lower will be the impact on the furnishings and structural system of the building. Loss of furnishings and/or the structural system causes direct economic setback (e.g., in terms of retrofitting or demolition and reconstruction costs) as well as indirect economic setback (due to loss of business or occupancy during the assessment and retrofitting operations). Further, the structural performance has a major role in ensuring safety of firefighters who typically enter a building after the fire has begun to affect the structural system. Collapse of parts or whole of the building during this period may cause loss of lives of the firefighters and may affect nearby dwellings as well. Fire hazards have been shown to cause monetary losses in the order of 1–2% of gross domestic product (GDP) in most countries (Hall, 2014).

Fire accidents and losses in the United States are recorded and reported annually by the National Fire Protection Association. According to NFPA (2020), fire departments reported 1.4 million fire accidents in 2020 in the United States. These fires caused 3,500 civilian fire deaths, 15,200 civilian fire injuries, and an estimated property loss of US$ 21.9 billion. Though fire accidents in or on built structures constituted 35% of the total fire accidents, the life losses in such accidents constituted a major portion (78%) of the total fire deaths and 86% of the total fire injuries. The direct property damages due of structural fires was US$ 12.9 billion (55%). Only a quarter (26%) of the fire accidents were residential which constituted about three-quarters (76%) of the total fire deaths. The United Kingdom recorded 153,278 fires, 221 fire-related fatalities, and 6,567 injuries in 2020. A total of 27,482 fires were recorded in dwellings and 755 in high-rise buildings of the total fire. Of the total fatalities and injuries, 176 fatalities and 5,051 injuries occurred in dwelling fires (cityfire.co.uk).

The magnitude of fire hazards has posed severe problems for emerging economies also. China records about 252,000 fire accidents in the year 2020. Of the total fires, 6,987 fire incidents were recorded in high-rise structures. These fire incidents caused 1,183 deaths and incurred a direct property loss of 4.01 billion yuan (US$ 620.59 million; english.www.gov.cn). Russia recorded around 144,199 fire accidents with 7,913 deaths and 9,650 injuries in 2018 (accessengineeringlibrary.com).

The fire problems pose severe threats to the economic development of India as well. According to the Indian Risk Survey (IRS) 2021 report (pinkerton.com), fire has been ranked fourth in the risk ranking (first is natural hazards, second is cyber security, and third is intellectual property theft) of operations in India. According to the accidental deaths and suicide in India (ADSI) report of the National Crime Records Bureau (NCRB) 2020 (ncrb.gov.in), there were 9,329 reported cases of fire which caused 9,110 deaths and 468 injuries. Out of these, 5,248 (about 57%) of the deaths were reported in residential buildings. It is worth noting that the number of deaths-to-fire ratio of the United States was 0.0025, the United Kingdom was 0.0014, China was 0.0047, Russia was 0.0549%, and India was 0.98%. India ranks 54 among 183 countries with a fire death rate of 3.65 per 100,000 people. While the method of recording accidents and losses vary from country to country, these statistics do provide overall trends and indications. Figure 3.1 represents statistical data of the fire deaths in some countries from 1959 to 2018 (ctif.org).

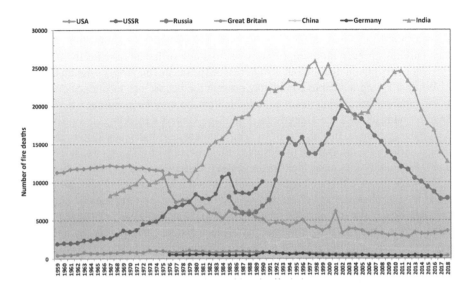

FIGURE 3.1 Trends of fire deaths (ctif.org).

The global fire statistics show a significant loss in terms of life, property, and economy caused by fire hazards in most countries. This discussion also shows that fire is a universal problem. Further, structural fires, specifically the residential building fires, represented the bulk of the reported fires because of various ignition sources such as cooking appliances, heating/cooling equipment, and electrical wiring. Also, such buildings contain a significant number of combustible items that constitute the fuel and usually untrained occupants. Fire accident studies have also shown that neither all fires grew to full size nor fully developed fires result in major structural damage. The adverse effect of fire on structural systems is quantified using the Poisson distribution approach to evaluate the fire breakout in a structure and fire-induced structural collapse. Some recent studies reported that the annual probability of fire breakout in a building is about 29.5%, and the probability of fire-induced structural collapse is about 12.1% (Lin, 2005; Naser and Kodur, 2015).

It is observed from the above figure that the annual probabilities of fire are frequent to almost every built infrastructure facility, when compared to other natural hazards such as earthquakes; latter has a very low probability of occurrence. Therefore, it is vital to design every structure considering the fire loading irrespective of its location. With the development of technology and rapid urbanization, an increase in the number of high-rise buildings is observed in developing countries. Among all the constructed infrastructural facilities, the probability of fire accidents is highest in buildings because of commonly present ignition sources, a significant number of combustibles as fuel load, and the presence of generally untrained occupants. Compartment fires are more dangerous than open fires from a life safety perspective because occupants may get trapped inside the buildings. Most high-rise buildings usually have a stay-put policy, which requires all the occupants to move to certain designated areas within the building (e.g., their apartments or a refuge area) and wait until they are evacuated by trained professionals. Structural fire safety of such buildings is even more important so as to allow sufficient time for the firefighters to not only put out the fire but also to rescue the occupants from designated areas.

While it is impractical to completely prevent the occurrence of fire accidents due to the inherent uncertainty (as also indicated by the very large number of reported fires in countries like the United States and the United Kingdom), it is possible to design the systems so that effects of such an accident (e.g., the death-to-fire ratios discussed earlier) are minimized by ensuring the provision of effective fire safety measures. Fire safety measures include prevention of ignition and growth of fire at the initial stage and fire resistance and firefighting at the later stages of fire and as such entail the use of active as well as passive systems. Active systems include automatic fire suppression systems such as sprinklers as well as other associated systems such as fire alarm systems and public announcement systems. These play an important role in the early stages of a fire. It is ideal to complete the evacuation of a building at the onset of a fire and to extinguish the fire while it is still small. This minimizes the risk to life, structure, and property. If the fire is not controlled in its incipient stage, compartment fires lead to a condition called flashover, discussed later in this chapter, and the size of the fire increases significantly. At this stage, it is practically impossible for anyone to survive and the structural system begins to get affected.

In post-flashover scenario, structural fire performance becomes important. Inherent resistance of structural members to fire (e.g., of reinforced concrete members) and provision of additional fire protection (e.g., board protection for steel members) constitute the passive fire protection system of a building. The efficacy of these systems is important not only to ensure structural fire safety but also firefighter safety, as discussed earlier. Another objective of passive fire protection is to arrest the spread of fire (through compartmentation) and to reduce the extent of damage to the main structural system. For instance, a good fire protection system of a steel structure will get sacrificed during a fire but will prevent permanent deformations in the main structural system so that during the retrofitting operations, only the fire protection system will be required to be replaced and not the main structural system. This not only helps in reducing direct financial losses but also in reducing indirect losses, as the downtime of the building is lower if structural retrofitting is not involved. Finally, fire protection engineering requires a multidisciplinary approach where the building is to be considered as a system and individual components cannot be considered in isolation.

3.2 DEVELOPMENT OF A FIRE

Fire is a dynamic phenomenon affected by factors such as type, amount, and characteristics of fuel, possible ignition sources and their strength, compartment geometry, ventilation conditions, and fire protection systems. It is necessary to understand and quantify the fire development and its behavior to lay out different strategies and provisions of fire safety measures. A fire is characterized by two characteristics: its heat release rate (HRR)-vs.-time behavior and its temperature-vs.-time behavior. While HRR-time behavior indicates the energy content and release rate of a

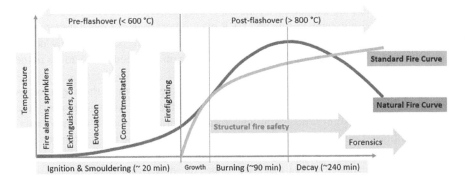

FIGURE 3.2 Time-temperature curve during the development, sustenance, and decay of a compartment fire.

fire, temperature-time behavior captures the variation of temperature throughout the duration of a fire. Quantification of both these behaviors is discussed later in this chapter. For design scenarios, a temperature-time curve is required, and all the fire ratings are also derived from such curves. Heat transfer principles can be used to generate temperature-time curves given the HRR-time curves and the compartment geometries and materials.

A typical temperature-time curve to be expected during an unsuppressed compartment fire is shown in Figure 3.2. A typical HRR-time curve is qualitatively similar to the temperature-time curve. Initially, the ignition happens at a certain portion of the room (e.g., due to a short circuit or due to a small open flame), which is called the incipient stage of fire. During this time, the compartment has sufficient supply of oxygen, and the size of fire grows as per the availability of fuel (combustible items). Usually, if combustible items are kept far from each other, chances of this local fire to spread to the nearby combustible items are small. If the initial combustible item has large energy content and proximity of other combustible items is good, the fire enters the growth stage. During this stage, the temperature (and the HRR) gradually rises. Production and acculumlation of hot gases and soot begins at the ceiling level of the compartment. Egress operations, initial firefighting operations (e.g., through activation of automatic sprinkler, use of portable extinguishers by occupants), and call to the nearest fire station should ideally be completed during this time.

If the initial measures work well, the fire is not expected to grow beyond this stage and is unlikely to cause any loss of life; it may cause minor economic losses though. The maximum temperatures seen by the surroundings up to this stage are in the range of 300°C, and there is no threat to the structural system of the building. On the other hand, if the fire continues to grow, the ceiling of the compartment begins a large reservoir of hot gases which begin to radiate heat back to the combustibles kept in the entire room. Radiation, being the fastest mode of heat transfer, heats the entire compartment simultaneously which leads to production of a large amount of combustible gases (these are typically the volatile organic content of items like furniture and plastics). This, in turn, enriches the local fuel-air ratio and in the presence of an open flame, the entire compartment catches fire almost instantaneously – an event that is called flashover. In the temperature-time curve and the HRR-time curve, flashover is seen as a sharp rise in temperature and HRR over a short span of time. Temperatures quickly reach in the range of 700°C. At this stage, the fire becomes fully developed and becomes ventilation-controlled, depending on the extent of openings in the compartment.

Due to being ventilation controlled at this stage, the growth of the fire subsides and it reaches the steady burning stage during which the average temperature remains almost constant. The magnitude of the temperature depends on the ventilation conditions. Lower the availability of oxygen, lower will be the fire temperature (and higher will be the amount of carbon monoxide produced due to incomplete combustion). During this post-flashover stage, the effectiveness of the structural and compartmentation elements of a building becomes important. In fact, the standard fire curve (discussed later) forms the basis of fire ratings of different components of a system and emulates the flashover and post-flashover stages of a real fire. Firefighting operations usually begin during this stage of the fire. Thus, the structural members, in addition to withstanding high temperatures, many times face sudden quenching due to impingement of water streams originating from firefighting operations. Certain testing standard requires building components to be tested against a hose stream after a pre-defined standard fire exposure. (One such example is the UL 10B test for fire door assemblies – an important part of the compartmentation strategy of a building.)

Burning continues until the fuel gets exhausted or the available oxygen gets depleted. Next, burning in the room gradually declines, represented by the decay stage in the temperature-time (and the HRR-time) graphs. During this stage, the burning rate is again governed by fuel. Sectional temperatures of the structural member continue to rise because of the lag effect and heat transfer from the surrounding hot air. It is necessary to ensure the stability and integrity of the entire structural system in this stage as well. Failure of structural fire protection systems or firefighting can cause an uncontrolled fire that can spread to the adjacent rooms, other stories, and even to neighbouring buildings depending on the layout of the building on fire. Fire can spread at various levels through flames, radiation or hot gases penetrating doors, windows, and cavities of the compartment.

In the early stage of fire development, the fire spread within the room depends mainly on the HRR of the initially burning object, and surrounding objects will get ignited by the flame impingement or by the radiant heat transfer. In later stages, the movement of hot gases under the ceiling can cause fire to spread in other parts of the room. Combustible lining materials on walls and ceilings can cause a rapid flame spread. Lining materials and various other combustible materials such as furniture items present in a room are characterized by properties such as ignitability, HRR, heat of combustion, and amount of smoke produced. These are often called 'reaction to fire' properties. Most of the countries have prescriptive codes for limiting these properties of lining materials depending upon the function of the building. Plastic or synthetic materials provide the least safety, whereas non-combustible materials like paper-faced gypsum plaster provide maximum safety in ignition and fire spread. Wood-based materials are moderately safe, and their resistance properties can be enhanced by using special paints or pressure treatment.

Such class of materials that are used to seal the gaps between different parts of a building are generically referred to as firestops. Depending on the area of application, they derive specific nomenclature. For instance, firestops that seal the gaps between the edge of the slab and façade systems are called perimeter fire barriers (or edge of the slab fire barriers); those used in holes used for crossing of the services (such as plumbing and electrical conduits) are called through-penetration fire barriers. These are typically made of insulation materials like stone wool and intumescent material.

The upper layer of hot gases along the ceiling can travel to the adjacent rooms and corridors through various openings and cause the heating of the succeeding areas, leading to the rapid-fire spread. The spread of toxic gases and smoke significantly contributes to fire deaths. The movement of the hot gases and the fire depends much upon the structure's geometry, as shown in Figure 3.3. Various openings like doors, concealed spaces, and penetrations for various fittings through fire-resisting walls can provide a path for the movement of fire and toxic gases. Automatic devices that close the door when the fire is detected are very effective. Doors passing through fire barriers (i.e., walls) should be designed for fire containment. Smoke control strips or intumescent paints that swell on heating effectively prevent the fire and smoke spread through gaps around the door.

Concealed spaces penetrating through fire-resisting walls are the most dangerous paths and create hazardous situations. Such spaces should be sealed with fire stopping materials. Other openings such as penetrations for the fittings and the construction joints should also be sealed with fire stopping materials. Various fire-stopping materials include mineral wool, gypsum board, metal brackets, fire-resisting putty, and intumescent collars. Fire-resisting walls must be extended above the suspended ceilings to the floor or roof above to ensure fire containment.

Fire can spread through various internal and external paths to other storeys, as shown in Figure 3.4. Internal paths include vertical shafts or stairways, interconnected vertical and horizontal concealed spaces, service ducts, gaps at the junction of floor and exterior wall or the failure of floor/ceiling assembly. Vertical openings and gaps must be entirely sealed with fire stoppings, as shown in Figure 3.5. Various vertical services can be enclosed in protected ducts or have fire stoppings at each floor level. External paths include combustible claddings or exterior windows. Small windows with horizontal apron projected above window openings can help control fire spread at higher storeys.

FIGURE 3.3 Spread of fire to adjacent rooms.

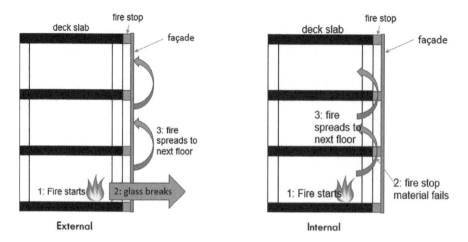

FIGURE 3.4   Spread of fire to different floors.

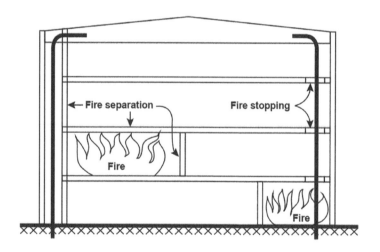

FIGURE 3.5   Use of firetops to ensure compartmentation.

In buildings with facades, two main fire spread mechanisms work: the first is the internal fire spread caused due to failure (or absence) of firestops between the edge of the slab and the façade mullions and transoms, and the second is the external fire spread, also called the leap-frog mechanism, in which the façade gives way to hot gases externally and the flames reach out to façade components of higher floors. The role of the spandrel area is critical in such scenarios. The choice of façade materials also plays a significant role. Readers are referred to Srivastava and Gandhi (2021) and the references therein for a detailed discussion on such mechanisms with respect to combustible façade systems.

3.3 FIRE SAFETY

Fire is a probabilistic event that can occur anytime in the structure's service life, irrespective of its location. A fully grown fire poses a serious risk to life safety and subjects the structure to resist the most severe conditions during its service life. Therefore, various strategies to reduce the effect of fire must be in place to control its effect on occupants, structure, and surrounding environment.

3.3.1 OBJECTIVES OF FIRE SAFETY

The primary goal of fire protection is to limit the probability of life, property, and environmental losses to acceptable levels. Depending upon the type of building and its occupancy, the balance between life safety and property protection varies. It can be done by laying out specific objectives that form the basis of fire protection systems and strategies.

3.3.1.1 Life Safety

Life safety is the most important objective for the safe escape of the people trapped in the building, especially in high-rises, in a fire. This objective ensures the life safety of the occupants within the building, the firefighters, by-passers, and the occupants of adjacent buildings. To achieve this, it is necessary to alert the people and provide suitable escape paths not affected by fire and smoke. In some buildings like hospitals, it is also necessary to provide safe places for the people unable to escape.

3.3.1.2 Property Protection

Event of fire not only costs the property losses, but it also involves the indirect loss that occurs because of interruption to businesses. The second objective of property protection is ensured by protecting the structural and non-structural members of the building and adjacent buildings and ensuring their functional continuity. The stability and integrity of the building fabric are important for fire containment and minimizing property losses.

3.3.1.3 Environment Protection

Gaseous pollutants such as toxic gases and smoke, chemical agents used in fire extinguishers, pollute the surrounding air and water bodies. The third objective is to reduce the environmental impact of fire which is achieved by extinguishing the fire at an incipient stage.

  All the above objectives are achieved by extinguishing the fire before it grows large, and it is accomplished most effectively with the help of automatic sprinkler systems.

3.3.2 SCENARIO ANALYSIS

Fire safety is a complex phenomenon involving a large number of interacting variables. Therefore, it is challenging to visualize fire safety without a conceptual framework. One of the frameworks is scenario analysis, often used in fire engineering design, which is shown in Figure 3.6 (Buchanan, 2001).

Another framework to visualize fire safety is the Fire Safety Concepts Tree developed by the National Fire Protection Association (NFPA, 1997). An overview of Fire Safety Concepts Tree is represented in Figure 3.7. The following paragraphs provide a brief explanation of the tree:

- Line 2 of the tree suggests two alternate ways for fire safety: either the ignition can be prevented, or the impact of fire must be managed. In reality, there will always be a number of unplanned ignitions, but fire prevention programmes can reduce their probability.
- Line 3 of the tree divides the management of fire impact into two parts, i.e., managing the fire itself, or managing the exposed persons and property.

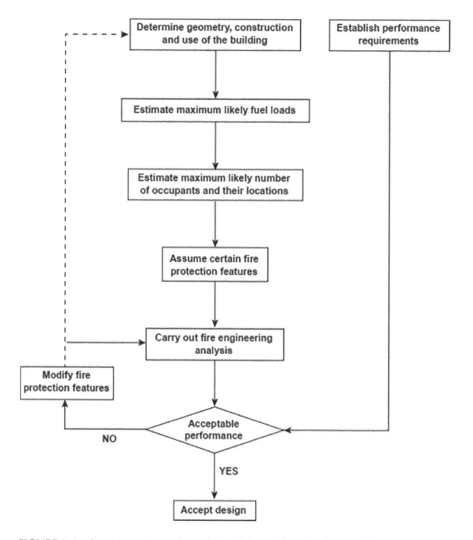

FIGURE 3.6 Overview of scenario analysis. (Adapted from Buchanan, 2001.)

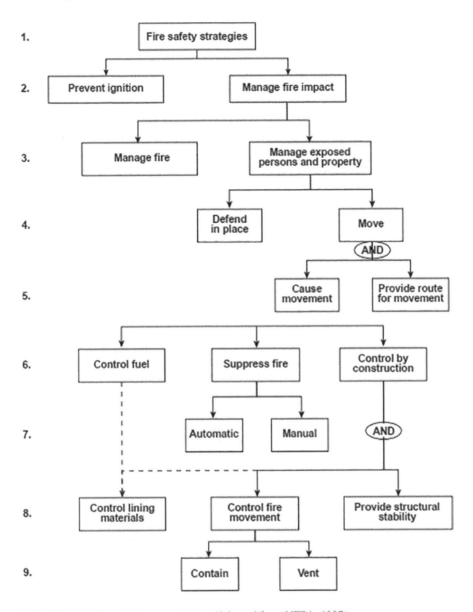

FIGURE 3.7   Fire safety concepts tree. (Adapted from NFPA, 1997.)

- Line 4 suggests the management of persons and property in two ways: either by defending them in place or by moving them outside the building. Generally, the exposed property is defended in place, whereas active persons move outside the buildings. There must be an intermediate refuge place for large buildings where the incapacitated persons are protected.

- Line 5 suggests that to move persons out of the building, it is necessary to alert them and provide them with a suitable egress path for their safe movement outside the building.
- Line 6 suggests three options for managing fire: controlling the fuel source, fire suppression and control by construction. The fuel source is controlled by limiting the amount of fuel and the geometry.
- Line 7 suggests that fire suppression is done automatically or manually, depending on the stage of fire.
- Line 8 suggests the measures for controlling the fire by construction, which is the scope of the book. It suggests that to control the fire by construction, it is necessary to control the fire movement and provide structural stability. On line 8, control of lining materials is connected by a dotted line with two options because limiting the fuel in combustible linings is a part of controlling the fuel source, and the selection and installation of linings is a part of the construction process.

The provision of structural stability is important for protecting people and property at the time of the fire. It also ensures that the structure will not collapse during the fire and ensure their functional continuity with minor repairs. Line 9 suggests the measures to control the fire movement either by containing the fire or venting it outside. Venting is a useful strategy to reduce the impact of fire in single-storey buildings or at the top storey of tall buildings. It is done by an active system of mechanically operated vents or by passive systems like melting plastic skylights. Venting can increase the local severity of fire because of increased ventilation but will reduce the overall thermal impact on the structure. Containment of fire is achieved by passive systems such as walls and floors. Fire spread to the adjacent buildings is also prevented by limiting the size and proper designing of openings in the exterior walls.

3.3.3 FIRE SAFETY MEASURES

The objectives discussed earlier are achieved by providing appropriate and well-designed fire protection features in the buildings. These features tend to minimize ignition, control and extinguish fires, alert the occupants, facilitate their safe evacuation, and prevent the collapse of the structure. Fire safety features are broadly divided into two categories: active fire protection systems and passive fire protection systems. These are explained in Chapter 4.

Structural integrity maintains the load-bearing capacity and stability of structural members in fire breakout. It is achieved by providing fire resistance in terms of structural fire resistance. When structural members and systems are exposed to fire, they experience a rise in their cross-sectional temperatures, which cause the deterioration in their mechanical properties and eventually cause their failure by the partial or complete collapse of structural systems. The constituent material of the structural member decides the type of deterioration. Such fire resistance systems minimize property loss. Therefore, structural members are designed to resist high temperatures with minimal strength degradation, and it depends upon the materials and structural system type used in construction. Thus, passive systems are effective in controlling the later stages of fire (fully developed fires), as structural and non-structural systems are designed to resist high temperatures.

3.3.4 STRUCTURAL FIRE SAFETY

Structural fire safety is most important in the construction of high-rise buildings from the life safety consideration. Various high-rise buildings have a stay-put policy to protect the people when they are not in the area directly affected by the fire. It suggests that people should stay in their respective flats unless the fire is not in their flat or the corridor outside their flat. Such policies are effective when the individual built flats are designed for fire compartmentation. But such policies can be disastrous in case of inefficient structural fire safety measures. One such example is the Grenfell tower fire in London. Initially built with fire safety features back in 1974, the tower went under major renovation between 2012 and 2016, which undermined the fire protection features. As a result, the fire spread quickly to the higher stories. The stay-put policy meant to safeguard people against fire trapped them inside their homes, which eventually caused 79 deaths, over 70 injuries and significant property loss.

Effective passive fire protection systems are also essential when the active fire protection systems may be absent or dysfunctional, such as in the case of fire following an earthquake, fire during construction, or in the post-flashover phase of the fire for the protection of firefighters. Chances of ignition are high in a post-earthquake scenario because of toppled furniture, movement of hot equipment, or electrical malfunctioning. Therefore, greater attention is required for passive protection and structural fire resistance in seismic regions than non-seismic regions.

Modern buildings have lower fire resistance because of various structural demands such as leaner structural members and open-concept architectural demands. Such demands are satisfied using high-performance concrete, high-strength steel, and engineered wood that provide high mechanical strength in ambient conditions. But these new construction materials possess lower fire resistance. They are prone to fire-induced spalling in concrete, which further reduces the leaner section, local buckling in case of high-strength steel, and delamination of wood at relatively lower temperatures. Further, the ever-increasing fuel load because of the increased use of plastics in furniture and interiors can cause more severe fires than experienced with cellulosic materials that dominated the buildings two decades ago has alleviated the problems. Therefore, the currently used prescriptive methods based on conventional materials and standard fire scenarios require rational extrapolation and adjustment for their applicability to account for realistic fire scenarios in evaluating the fire resistance of modern structures.

Fire resistance often refers to passive fire protection systems, and active fire protection systems are required to design fire protection strategies. Fire resistance is defined as an ability of a structural or non-structural member to maintain its load-bearing capacity, integrity, and insulation capacity against fire during its entire duration. Structural components are designed to withstand the service loads, and partitioning components are designed to prevent fire spread to adjoining spaces.

Fire resistance is specified in terms of fire ratings (measured in duration) in three domains: load, insulation, and integrity. Fire ratings represent the duration for which a structural or non-structural member can resist their exposure to the representative standard fire along with maintaining its load-bearing capacity, integrity, and/or insulation criteria. It depends upon the type of occupancy, compartment area, and the height of the building. Fire ratings are usually 30 min, 1 h, $1\frac{1}{2}$ h, 2 h, 3 h, or 4 h and the values are rounded off to the nearest 30 min.

A structural member is designed such that the actual fire resistance rating of the member is higher than the required fire resistance rating according to the fire design demand.

The fire design time depends on the importance of the building and the consequences of structural collapse or fire spread. The designer selects fire design time from the following:

1. The time required for the safe escape of the occupants from the building
2. The time required by firefighters to carry out rescue activities
3. The time required by firefighters to contain the fire
4. The time required for complete burnout of the fire compartment with no intervention

The actual fire resistance rating of a structural member is evaluated either by using prescriptive based approaches or rational calculation methods.

Prescriptive-based approaches are simple as they evaluate the fire resistance based on tabulated data, simplified rules, or calculation methods derived from standard fire tests for a given member. The analysis is carried out at the sectional or member level and not at the entire system using these approaches because they do not account for design variables such as fire loading and end restrains, and the realistic failure modes. Such approaches can lead to unrealistic fire performance assessment when the real fire growth characteristics are not similar to a standard fire.

Rational methods evaluate the fire resistance of the structures by considering the actual conditions present during a fire scenario using engineering-based calculation methods. They can be applied at sectional, member, and system levels because rational approaches account for critical design factors such as material, geometric properties of a member, fire loading, end restraints, and boundary conditions. A number of rational methods have been developed with technological development in the last few decades. However, they are still evolving, and their application is at an incipient stage as they require validation and benchmarking.

3.4   CODES AND STANDARDS

Fire design specifications and the guidelines to achieve them and fire safety requirements are mentioned in the building codes of each country. These codes specify the minimum requirements for fire safety features in the building to achieve the desired level of fire safety. Codes from various countries specify the fire rating of various structural members in terms of time depending on the occupancy type, structure height, and maximum compartment area. A standard test is established to achieve uniformity in the fire ratings specified by different countries. Fire test standards such as ISO 834 (ISO, 2012), ASTM E119 (ASTM, 2020), UL 263 (UL, 2011), and IS 3809 (BIS, 1979) include the specifications for evaluating the fire ratings of structural members by testing. Fire resistance rating is defined by the duration for which a structural member can resist a standard fire exposure without reaching the limit state of failure, as specified in the test standard. Fire resistance ratings of structural members are also evaluated by engineering approaches provided in the standards such as Eurocodes (CEN, 2004, 2005) and ASCE 29 (ASCE, 2005). A flowchart for such calculations is given in Figure 3.8, and a brief discussion on the key elements follows.

Fire model forms the basis for understanding the fire development and behavior within the compartment and quantifies it by time-temperature curves. It involves the concepts of fire load, HRR, the effect of geometry, and ventilation required as inputs in fire models. The output of the fire model is the estimation of the HRR-time and/or temperature-time curve which can be a standard, parametric, or measured/simulated real fire curve. Concept of equivalent fire severity among the different types of fire curves is also important.

Heat transfer model covers the interaction between the fire and the structure. It is used to compute the temperature rise on exposed surfaces and temperature gradients within a structural member. Fire resistance of a member in terms of either load capacity or fire containment depends upon its internal temperatures. The rise in temperature of the materials exposed to fire depends on radiation and convection at the surface (fire to structure heat transfer) and on thermal conduction within the member. Heat transfer depends on many parameters such as the geometry of the member,

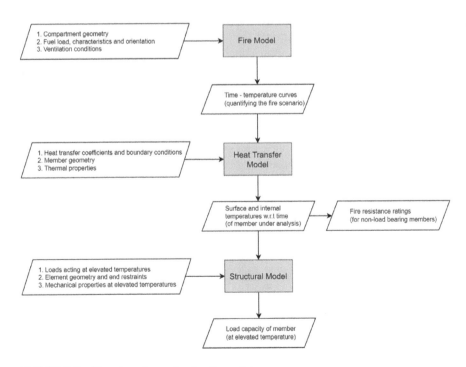

FIGURE 3.8 Flowchart for evaluating fire resistance.

convective and radiative heat transfer coefficients at boundaries, and thermal properties of the materials. Fire resistance of non-load bearing members and assemblies is determined based on their ability to contain a fire (a key idea of compartmentation). The output from the heat transfer model is directly used in evaluating the thermal resistance of such members in terms of time when a critical temperature or condition is reached on the unexposed side. Fire resistance ratings of insulation, as discussed later, is computed directly as an outcome of the heat transfer model.

Once the temperatures at surfaces and within the members are obtained from the heat transfer model, the strength of individual sections, assemblies, and sub-assemblies can be calculated. Simple hand calculations are possible for individual members by considering the reduction in the strength of materials at elevated temperatures and involve sectional analysis; such calculations form the basis of most of the building codes. For the analysis of sub-assemblies and systems, computer-based structural analysis models are required, which consider the effects of end-restrains, thermal expansion, large deformations, and non-linear temperature-dependent material properties. Many building codes provide guidance on such advanced structural analysis.

3.5 FIRE BEHAVIOR

This section reviews the basic concepts of combustion of fuel in building fires and provides a brief idea about calorimetry used to determine the burning behavior of the fuel materials. It also discusses the concepts related to the quantification of the fire load. The chapter reviews the fire development process within the compartment and various empirical fire development models that consider fire load and ventilation conditions. Additionally, temperature progression at different stages of fire in terms of standard and parametric fire temperature-time curves is also discussed.

Fire is a combustion phenomenon requiring three essential components to initiate fuel, oxygen, and heat, and is represented in the form of a fire triangle as shown in Figure 3.9. The presence of all the three components in the right proportion can lead to the fire, and removal of any one component can extinguish the fire. The general chemical reaction denoting complete combustion during a fire is given in Eq. (3.1).

$$\text{Fuel} + \text{Oxygen} \ \rightarrow \ CO_2 + H_2O + \text{Heat} \tag{3.1}$$

FIGURE 3.9   Fire triangle.

Equation (3.1) shows that fire is an exothermic reaction and it produces heat, carbon dioxide, and water vapour. In case of incomplete combustion (usually the case with most building fires), a large amount of carbon monoxide and soot particles is also produced.

Fire in a compartment is initiated by igniting fuel in the presence of oxygen (natural air). Once ignition takes place, a self-sustained fire can form if there is sufficient supply of fuel and oxygen and the geometries and placement of combustibles are favourable. This process was discussed earlier in the development of a fire. The combustion process involves two reactions – thermal decomposition of the object to produce fuel vapours (typically called pyrolysis; it is endothermic in nature) and oxidation reaction that burns the fuel and further speeds up the combustion process (the combustion; it is exothermic in nature). The fire will self-extinguish if the entire fuel is consumed or the heat is dissipated, causing the temperature to drop below the ignition temperature of surrounding materials or the entire room is filled with smoke creating oxygen deficiency.

3.5.1 FUELS

Any material in solid, liquid, or gaseous form having carbon in its composition is called fuel. There can be non-carbon–containing fuels as well; for instance, certain pyrotechnic materials and strong oxidizers may serve as fuels. Aluminum, a commonly used metal, can also burn at relatively modest temperatures (it melts at 650°C). These materials include organic material derived from trees or animals (naturally occurring fuels) and petrochemicals (synthetic fuels). Materials like cotton, jute, straw, wood, and many others are derived from trees. Animal-based organic materials include wool, leather, and various food products. Petrochemicals include various plastic materials, liquid fuels, and gaseous fuels. Many of these materials are used in building components, lining materials, or furniture that constitutes the potential fuel. All the fuel materials are hydrocarbons, and their molecule primarily consists of carbon and hydrogen along with oxygen, nitrogen, or any other in some cases. Each fuel material has a specific level of thermal energy stored depending upon its composition and is measured as a calorific value.

3.5.2 CALORIFIC VALUE

The total amount of energy released during the complete combustion of a unit mass of fuel is called the calorific value of heat of combustion of the substance. It is measured in J/kg. Most fuels (solid, liquid, or gas) have calorific values between 15 and 50 MJ/kg. The calorific values of commonly used materials in buildings are reproduced in Table 3.1. Under normal conditions, some combustible materials such as wood can contain moisture, reducing their calorific value. Therefore, the amount of energy released by the wet sample is lesser than the dry sample and is calculated as

$$H_u = H_{u0}(1 - 0.01u) - 0.025u \qquad (3.2)$$

TABLE 3.1

Fire Load Densities for Different Occupancies

Code	Average FLED (MJ/m²)		
	Residential	Office	Hotel room
Eurocode 1 (CEN, 2002)	780	420	310
NBC Part IV (BIS, 2005) [a]	425	425	425
New Zealand code (MBIE, 2007)	400	800	—

[a] These values were mentioned as design values in the code. The recent update, NBC 2016, has removed these values and recommends consideration of actual scenario.

where u is the moisture content expressed as a percentage of dry weight, and H_{u0} is the net calorific value for dry material (MJ/kg).

The total energy in the fuel can be computed by knowing the mass of fuel and its calorific value, and it is the maximum energy that is released by burning the fuel.

$$E = M \times H_u \tag{3.3}$$

where M is the mass of combustible fuel, and H_u is the net calorific value.

3.5.3 HEAT RELEASE RATE

Fire load quantifies the total energy released by the burning of fuel, but it does not provide a complete sense of the severity of the fire. Time is an important parameter that provides a sense of the severity of the fire. Suppose two fire cases release the same amount of 100 MJ of energy. In the first case, the fire continues to burn for 10 min and releases total energy, whereas in the second case, the fire burns for 30 min and releases the total energy. The HRR is defined as the amount of energy E (in MJ) released in a certain time t (in seconds). It can also be defined as the rate of heat release with respect to time. Hence, the average HRR Q (in MW) is given by

$$Q = \frac{E}{t} \tag{3.4}$$

and a plot of the two such fire scenarios is shown in Figure 3.10.

It can be observed that the first case is more severe for the above scenario and will create a greater threat to structural safety. The HRR from a combustion reaction depends on the energy content and the time, which depends on the nature of the burning material, the size of the fire, and the amount of oxygen available. HRR curves are generally used for characterising the fire behavior of the individual fuel material using calorimetry-based experiments. The area of the HRR curve is equal to the total energy released during a fire.

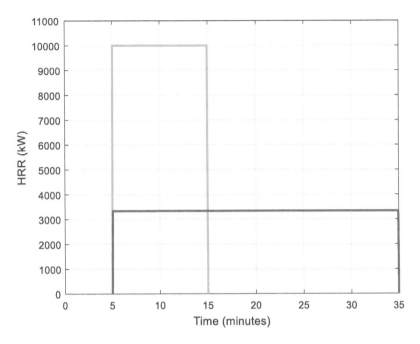

FIGURE 3.10   Average HRR for two fire scenarios releasing the same amount of total energy.

3.6   FIRE LOAD

The spatial distribution of combustibles within the room is also essential for characterising fire behavior. Consider the two cases with the same fire energy. In one case, the total energy is released in a room with a smaller floor area, and in another case, the total energy is released in a larger room with a larger floor area. The first case is more severe as the total energy is released within a small area. Therefore, fire load in a compartment is represented in terms of Fire Load Energy Density (FLED) and is expressed by

$$e_f = \frac{E}{A_f} \text{ or } e_t = \frac{E}{A_t} \tag{3.5}$$

where E is total fire load, e_f is the energy density per square meter of floor area (MJ/m^2 floor area), A_f is the floor area of the room, e_t is the energy density per square meter of total area (MJ/m^2 total area), and A_t is the total area (floor, ceiling and walls, including window openings) of the room. In Eq. (3.5), e_f is observed to be greater than e_t for a given room by the ratio A_t/A_f. Therefore, which fuel density is used depends on the situation, and a major error can occur if the distinction is not clear.

The fire load energy density is an important parameter for quantifying fire load in the buildings, and it depends on the type of occupancy of the buildings. Various countries have conducted extensive fuel load surveys to standardize the FLED for fire design in buildings of different occupancies. Typical values of FLED lies between

100 and 10,000 MJ/m² (with respect to floor area). Average fire load energy densities for different buildings from various codes are given in Table 3.1.

Design fire loads are determined as extreme values of likely fire scenarios similar to the design loads obtained for other extreme events such as earthquakes or wind. It includes both the fixed and movable fire loads. The design fire load should ideally have an exceedance probability of less than 10% for 50 years (typical service life of a building). The design fire load should be selected for safe design between 80th and 90th percentile of the surveyed fire load (the Eurocode suggests using the 80th percentile of the FLED as the design FLED value). For a coefficient of variation of 50%–80% of the average value, the 90th percentile value is 1.65 to 2.0 times the average values given in Table 3.1 (Buchanan and Abu, 2017).

3.7 FIRE INITIATION AND FLAME SPREAD

The fire, a combustion process is an exothermic reaction represented by (1). The stagewise involved process in combustion is discussed next.

3.7.1 PHASE CHANGE AND DECOMPOSITION

At room temperature, most of the fuels in the building are solid or liquid states, and some are in the gaseous state. Solid, liquid, or gaseous fuel follows a slightly different process for the initiation of fire. The gases mix readily with the air and burn directly. Whereas solids and liquids undergo a phase change to gas, their gaseous phase mixes with the air and then burns. Most of the liquids are converted to vapour on the application of heat by the process of evaporation. The thermal decomposition process converts some polymeric compounds to new volatile products. Solid fuels on heat application first melt into the liquid phase, and then liquid vaporize to the gaseous phase. Some solid fuels such as wood thermally decompose directly to the gaseous phase from the solid phase. This process is called pyrolysis.

3.7.2 FIRE INITIATION

Auto-ignition or piloted ignition of the fuel material is required to initiate fire. Piloted ignition occurs in the presence of a flame or sparks, whereas spontaneous ignition of gaseous fuel in the absence of an external source (flame or spark) is auto-ignition. In most cases, an external ignition source is required to ignite fuel. The possible ignition sources in the building include flaming sources (matches, candles, open fires), smouldering sources (incense stick, cigarettes), electrical sources (overheating, short-circuits), radiant heat sources (sunlight, heaters), mechanical sources (friction), lightening, and others. Wars and terrorism also cause fires in buildings. Various fire prevention strategies can control many possible ignition sources, but it is not possible to completely eliminate the fire.

The temperature and amount of heat required for ignition depend on the type, shape and size of the material, and the time of heat exposure. A competent ignition source provides sufficient heat to raise the temperature to the fire point within the expected exposure time. The ignition time depends upon the thermal inertia of the material, which is the product of thermal conductivity, density, and specific heat. A material with lower thermal inertia, such as polyurethane foam, heats more rapidly than the material with higher thermal inertia, such as wood, causing their early ignition.

Auto-ignition temperature (AIT) is the lowest temperature above which a material may not require any external ignition source for combustion. AIT is also referred to as self-ignition temperature, supplying the activation energy that is needed for combustion. The fire point is lower than the AIT at which the fluid can sustain fire when ignited by an external ignition source.

3.7.3 FLAME SPREAD

After ignition, the fire safety in the building depends significantly on the flame spread rate. Initially, fire spreads by the ignition of adjacent combustible materials by flames. The rate of flame spread depends on the rate of heating of the fuel ahead of the flame. The heating rate depends on the size and location of flame for radiative heat transfer, the direction of airflow for convective heat transfer, and the thermal properties and flammability of the fuel. Flame spread more rapidly in the direction of wind flow rather than the opposite direction. The flames also spread rapidly upward as they rapidly preheat the material ahead of the burning region. The fuel with low thermal inertia, such as foam, is more susceptible to ignition because of rapid temperature rise, causing rapid flame spread and fire growth.

Once the fire reaches an established burning stage, the flames are large enough to sustain the combustion reaction without any assistance from an external heat source. The burning is further driven by the heat of the flame that causes the production of volatile gases by heating the fuel to a sufficient temperature, which burns in a dynamic process, producing more volatiles and flames.

3.8 COMPARTMENT FIRES

A fire within any enclosed space with physical boundaries to air circulation is a compartment fire. Fire initiates in a compartment within a building, and then if not controlled, it can spread to various compartments within the building. Therefore, the study of compartment fire is essential to understand the effect of fire on the building. As discussed earlier, the fire development process in any compartment, regardless of its size, fuel type, or ventilation conditions, is grouped into four sequential stages: incipient, growth, fully developed fire, and decay. The incipient and growth stages are grouped under pre-flashover fire, the fully developed stage and decay stage are grouped under post-flashover fire, and the transition is defined as flashover. The fire behavior is well understood and quantified using the time-temperature (shown in Figure 3.2) or time-HRR curves.

At flashover, the local fire is transformed into a full compartment fire within a short time. Flashover primarily occurs because of radiative heat feedback from the layer of hot gases accumulated near the ceiling and is marked by a rapid increase in temperature and HRR. The term 'flashover' is typically used for compartment fires, and it occurs when the temperature in the compartment reaches around 600–700°C. The time to flashover is the available safe egress time for the occupants because conditions can become life-threatening at the flashover stage because of high temperature and toxic gases. The fire can spread through multiple compartments or floors at a fully developed stage. The temperature in the compartment is around 600–1,000°C. The fire continues to burn until the fuel is consumed or the oxygen runs out, after which the fire will start to decay slowly. It is the transition of fire from a fully developed stage to a decay stage. In post-flashover fire, if oxygen availability is sufficient, then the fire is fuel controlled or else, it is ventilation controlled. The subsequent sections discuss fire behavior and HRR calculations at each of the aforementioned stages.

3.8.1 PRE-FLASHOVER FIRES

Initially, in the pre-flashover stage, the combustion is restricted to a small place in the room. Therefore, the overall HRR and the temperature are smaller enough to be ignored for structural design considerations. The study of fire behavior is important from the occupants' life safety and safe evacuation perspective. Therefore, understanding fire development, HRR, temperature, and smoke movement are necessary for the design and timely activation of the active fire protection systems. An early-stage fire, when only one item is burning in a room with one door, is shown in Figure 3.11. Initially, the combustion utilizes the oxygen from the air in the room, but as the fire develops, it utilizes the air coming through the opening.

The fire plume acts as a pump, pulling the cool air inside and pushing the combustion products outside the room. The combustion products rise and get collected near the ceiling. This forms a hot upper layer of gases. The lower layer consists of incoming cooler air. As the fire grows, the energy released by fire increases, increasing the temperature, volume of gases in the upper layer, and compartment pressure. The increase in the pressure causes them to move down and outside the room through the openings. The pressure in the lower layer containing cool air is low, causing the inward movement of fresh air from outside. The hot and cool layer intersection has the neutral pressure, known as the neutral plane. As the fire plume reaches the ceiling, the hot gases flow radially outward underside the ceiling, called a ceiling jet. The flow of gases depends on the shape of the ceiling. The hot gases will activate the active fire protection systems such as sprinklers or heat detectors near the ceiling. Simple hand calculations or the computer models like zone models can be used to characterize the fire development in the pre-flashover stage. As discussed earlier, the pre-flashover stage is not critical for structural fire safety.

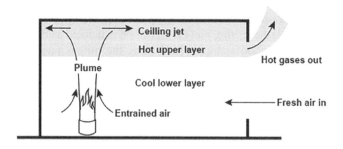

FIGURE 3.11   Early stages of a fire in a compartment.

3.8.2 t^2 FIRE MODEL

For design purposes, the HRR in the growth phase is often characterized by a parabolic curve such that the HRR is proportional to the time square and is defined as t^2 fires. On ignition, each fuel item in the compartment releases a certain amount of thermal energy (E) by burning within the time (t). This burning of the object can be characterized using the t^2 curve and is represented in Figure 3.12.

After ignition, the HRR continues to rise parabolically until the peak HRR (Q_p) is attained and then burns at a steady rate until the thermal energy contained in the fuel items is dissipated. The t^2 HRR is calculated as

$$Q = \left(\frac{t}{k}\right)^2 = (\alpha t)^2 \tag{3.6}$$

where Q is the HRR (MW), t is the time (s), k is the growth constant (s/MW$^{0.5}$), and α is the fire intensity coefficient (MW/s^2). The fire is classified as slow, medium, fast, and ultrafast based on the growth rate. The values of k are given in Table 3.2 for slow, medium, fast and ultrafast fire growth rate, and the corresponding HRR is shown in Figure 3.13. Values of t and Q_p are determined experimentally (Babrauskas and Grayson, 1992).

These t^2 design fire curves are often used as inputs in the fire models to predict the fire development in the compartments.

3.9   ENERGY AND BURNING TIME

Calculations for the energy and time of burning are discussed below regarding the t^2 fire shown in Figure 3.14. With the knowledge of peak HRR Q_p (MW) and the type of burning, the time t_1 (s) to reach the peak HRR is calculated as

$$t_1 = k\sqrt{Q_p} \tag{3.7}$$

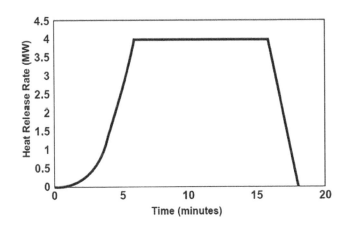

FIGURE 3.12   Typical t^2 fire for a single object. (t^2Model used in growth phase.)

TABLE 3.2

Fire Growth Rate Coefficient for t^2 Model

Fire Growth Rate	Growth Constant, $k \left(s/MW^{0.5} \right)$	Fire Intensity Coefficient, $\alpha \left(\times 10^{-6}\ MW/s^2 \right)$	Representative Real Fire Scenario
Slow	600	2.93	Densely packed wood items
Medium	300	1.17	Solid wood furniture, small amount of plastic
Fast	150	46.6	Furniture with moderate amount of plastic and/or upholstery
Ultrafast	75	187.4	Furniture with high amount of plastics and synthetic materials, and petrochemical products

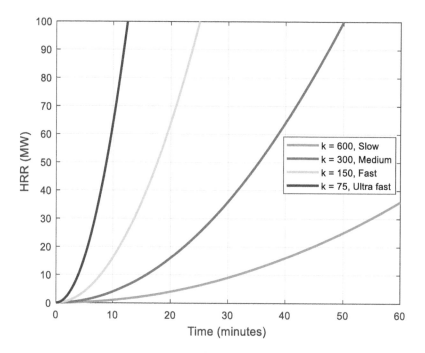

FIGURE 3.13 HRR for different types of t^2 fires.

The total energy released by the burning of the object is equal to the area under the curve of HRR vs. time. The energy E_1 (MJ) released during the growth phase under parabola is given by

$$E_1 = \frac{t_1 Q_p}{3} \tag{3.8}$$

where Q_p is the peak HRR (MW) of the object, t_1 is the time to reach the peak (s).

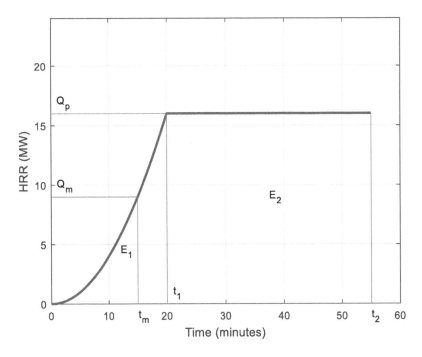

FIGURE 3.14 Calculation of energy from HRR.

If the total fuel has energy E is greater than E_1, then the remaining energy $E_2 = E - E_1$ is released during the steady burning phase. The total burning duration t_b (s) is given by

$$t_b = t_1 + \frac{E - E_1}{Q_p} \tag{3.9}$$

If the total fuel energy is less than E_1, the total fuel will be consumed in time t_m (s).

$$t_m = \left(3Ek^2\right)^{\frac{1}{2}} \tag{3.10}$$

And the burning rate is given by

$$Q_m = \left(t_m / k\right)^2 \tag{3.11}$$

3.9.1 Pre-Flashover Fire Calculations

The burning of each object in the compartment can be characterized using a t^2 fire curve. The fire spreads from one object to another by flame if objects are too close or by radiation. The ignition time of the other objects depends on the radiation intensity from the fire and the distance between them. By calculating the ignition time of the second object, the combined HRR by burning both the objects is obtained by adding

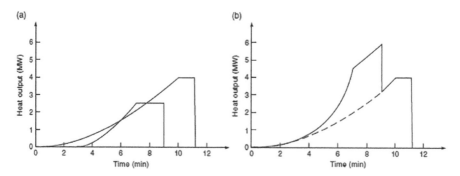

FIGURE 3.15 Combining t^2 fire HRR for two burning objects.

the HRR curve of both the objects at any time, representing their subsequent burning. For many objects present in the compartment, the resultant design fire can be approximated as t^2 fire for simplicity. The resultant HRR curve obtained by adding two HRR curves is shown in the Figure 3.15.

It is important to note that the design t^2 fires assume open-air burning (unlimited ventilation) conditions without any suppression. But these objects will burn differently in the compartment. The burning rate will be enhanced by radiative and convective heat transfer while limited by the available ventilation. These design fires of individual objects or their combined design fire burning rate can be used as inputs to room fire computer models to predict the actual fire development process. Simple hand calculations can be done for the room fires, where the peak heat rate (Q_p) should be a minimum of the total peak heat rate obtained by combining all fuels or the ventilation-controlled HRR discussed later.

3.10 FLASHOVER

Flashover is the transition from the localized fire to burning all the combustibles within the compartment. When the fire is allowed to grow without any intervention, it spreads to the surrounding combustibles, increasing the temperature in the compartment. When the temperature of the upper hot gas layer reaches a critical value of 600°C, resulting in a radiant heat flux of about 20 kW/m² at the floor level, all the exposed combustible items will begin to burn. This transition is a flashover. After flashover, the entire room is involved in fire and is termed fully developed or post-flashover.

Flashover occurs if certain conditions are fulfilled. They are as follows:

1. Sufficient availability of fuel
2. Sufficient availability of oxygen (ventilation conditions)
3. The ability of the compartment to trap hot gases (a compartment with significant openings in the ceiling cannot hold the required amount of hot gases)
4. The geometry of the compartment should be such that the radiant heat flux from hot gases should reach the other combustible items and increase their temperature to their critical ignition levels

It has been observed from various experiment fires that the flashover can only occur if the HRR from fires reaches a specific critical value based on the size of the ventilation opening. The critical value of HRR (Q_{fo}) for a room with a single window is empirically given by 'Thomas's flashover criterion' as follows:

$$Q_{fo} = 0.0078A_t + 0.378A_v\sqrt{H_v} \tag{3.12}$$

where A_t is the total internal surface area of the compartment (m^2), A_v is the area of the window opening (m^2), and H_v is the height of the window opening (m). Note that Eq. (3.12) excludes the dependence of flashover on factors such as fuel distribution, fuel type, and lining materials.

If the burning of all objects in the compartment is characterized using t^2, then the time to flashover can be roughly predicted using the critical HRR Q_{fo}:

$$t_{fo} = k\sqrt{Q_{fo}} \tag{3.13}$$

where Q_{fo} is the critical HRR for the given room (MW), t_{fo} is the time to flashover (s), and k is the growth constant (s/$\sqrt{\text{MW}}$).

3.11 POST-FLASHOVER FIRES

After flashover, high temperature and radiant heat flux situations prevail in the compartment in the fully developed fire stage. It can cause the production of a large amount of combustible gases by the pyrolysis of all the combustible surfaces, which eventually leads to their burning in the presence of sufficient ventilation (oxygen). The temperature rises and may attain the maximum temperature of over 1,000°C in the compartment. The flow of combustible gases and the air becomes turbulent. After attaining peak temperature, the temperature within the compartment will reduce as the rate of combustion decreases and the fire enters the decay stage. Based on compartment characteristics, the behavior of post-flashover fire is grouped into two groups: fuel-controlled fires or ventilation-controlled fires. Until the time large burning surface areas are involved in burning, the fire is ventilation controlled, and as the surface area reduces, the fire becomes fuel controlled.

The structural design of the compartment is tested as the structural members experience a high-temperature rise at this stage. The structural members are exposed to the most severe effects of fire in this stage that can cause the loss of structural capacity and integrity and even collapse. The information about the temperature in the room is the most important for the structural design. Factors such as the size and type of fuel materials, size of openings (ventilation condition), and the compartment boundaries (lining materials) decide the temperature decay rate in the compartment in the post-flashover fire stage.

3.12 VENTILATION-CONTROLLED FIRE

Many fire scenarios have limited ventilation in the compartment in the post-flashover stage leading to the ventilation-controlled fire. The rate of burning depends on the shape and size of the ventilation opening, which controls the amount of cold air entering the compartment and, correspondingly, the amount of hot gases leaving the compartment. In ventilation-controlled fire because of insufficient air in the compartment, flames extend outside the window to allow the complete combustion of combustible gases leading to the additional combustion by mixing unburned gas fuels with the outside air. It is conservatively assumed that the glass windows (except wired or fire-resistant glass) break and fall out because of the high temperature rise at the time of flashover. The fire continues to burn at a lower HRR for a long time if the glass does not fall out.

3.12.1 BURNING RATE

Based on many experiments, Kawagoe (1958) approximated the burning rate $\dot{m}$ (kg/s) of wood fuel for a room with a single opening as

$$\dot{m} = 0.092 A_v \sqrt{H_v} \tag{3.14}$$

where A_v is the area of the window opening (m²), and H_v is the height of the window opening (m). The corresponding ventilation-controlled HRR Q_{vent} (MW) for steady-state burning is obtained by

$$\dot{Q}_{vent} = \dot{m}\Delta H_c \tag{3.15}$$

where ΔH_c is the heat of combustion of fuel (MJ/kg).

If the total amount of fuel in terms of mass M_f (kg) or energy E (MJ) available in the room is known, then using burning rate or HRR, the duration of burning t_b (s) for the given room is also calculated as

$$t_b = \frac{M_f}{\dot{m}} = \frac{E}{Q_{vent}} \tag{3.16}$$

All the above calculations depend on the approximate relation of the burning rate given in Eq. (3.14) is widely used but may not always be accurate because of the complex fire growth characteristics and the underlying assumptions used to develop the principal relation. Various conditions such as non-uniform distribution of fuel items in the room or some portion of specific fuel type that may not be available for combustion can lead to inaccurate estimation of fire growth. Ventilation-controlled fire involves a complex interaction between the inflow of air to support combustion, the outflow of combustion gases and unburnt fuel from windows, the radiant heat flux on fuel, the pyrolysis rate, and the burning rate of gaseous fuel. These interactions depend upon the type of fuel, the shape of fuel (cribs or lining material), and the ventilation openings. Many studies show the empirical dependence of the ventilation-controlled burning rate on the term $A_v \sqrt{H_v}$, but some tests have also shown departure from Kawagoe's equation. Despite these approximations, Kawagoe's equation will form the basis of most post flashover fire calculations until further research is conducted.

Law (1983) proposed a more refined equation considering the floor shape of the compartment for the burning rate based on a large number of small-scale compartment experiments with wood cribs reported by Thomas and Heselden (1972):

$$\dot{m} = 0.18 A_v \sqrt{\frac{H_v W}{D}} \left(1 - e^{-0.036\Omega}\right) \qquad (3.17)$$

where $\Omega = \dfrac{A_t - A_v}{A_v \sqrt{H_v}}$, W is the compartment width (m), D is the compartment depth (m), and A_t is the total area of the internal surfaces of the compartment (m²). Eq. (3.17) is only applied directly to the compartments with windows on one wall because it is difficult to differentiate between W and D if there are windows on two or more walls. The burning rate calculated using Eq. (3.17) is given in Eurocode (CEN, 2002) to assess the flame height from the compartment window. For square compartment with ventilation factor of $A_v \sqrt{H_v} /A_t = 0.05$ ($\Omega = 20$), Law's equation (3.17) gave approximately the same burning rate as Kawagoe's equation (3.14). For wider shallower compartments with smaller openings, the burning rate is greater than Eq. (3.14).

3.12.2   Ventilation Factor

The amount of ventilation in a fire compartment is described using ventilation factor (opening factor) F_v (m$^{0.5}$):

$$F_v = A_v \sqrt{H_v} /A_t \qquad (3.18)$$

where A_v is the area of the window opening (m²), H_v is the height of the window opening (m), and A_t is the total area of the internal surfaces of the compartment (m²). If the acceleration of gravity g is introduced in the above factor, the unit of term $A_v \sqrt{g H_v} /A_t$ will be m/s, related to the gas flow velocity from the openings.

3.12.3   Effects of Multiple Openings

For more than one opening, the above equations can be used by modifying A_v as the total area of all openings and H_v as the average height of all doors and windows. The underlying assumption for openings on several walls is that the airflow is similar at all openings, and there is no strong wind that creates cross-flow through the room.

For the compartment shown in Figure 3.16, the average height of the openings H_v, the area of openings A_v, and the total internal area A_t are calculated as

$$H_v = (A_1 H_1 + A_2 H_2 + ...)/A_v$$

$$A_v = A_1 + A_2 + ... = B_1 \ H_1 + B_2 \ H_2 + ...$$

$$A_t = 2(l_1 l_2 + l_1 H_r + l_2 H_r) \qquad (3.19)$$

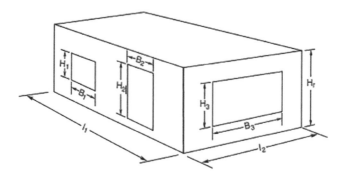

FIGURE 3.16 Calculation of ventilation factor for more than one opening.

where B_i and H_i are the breadth and height of the windows, l_1 and l_2 are the floor plan dimensions, and H_r is the compartment height.

3.13 FUEL-CONTROLLED FIRE

Fires are fuel-controlled in large, well-ventilated compartments containing fuel items with a limited area of combustible surfaces. In the decay period, mostly the fires are fuel-controlled. The burning rate is similar to the fuel item burning in the open air, enhanced by the radiant heat from the upper hot gas layer near the ceiling and hot walls. The average HRR Q_{fuel} can be calculated for a fuel-controlled fire with the knowledge of total fuel load and burning duration. Law (1983) concluded from many experimental fires that a typical domestic furniture fire has a burning duration t_b of around 20 min (1,200 s). So, for residential fires, the average HRR Q_{fuel} (MW) for fuel-controlled fire can be calculated as

$$Q_{fuel} = \frac{E}{1,200} \tag{3.20}$$

where E is the total amount of fuel load (MJ) in the compartment

In most cases, the duration of the fire is not known. For the fuel-controlled fires, when the burning is controlled by the available surface area of fuel items, Drysdale (1998) represented the average HRR Q_{fuel} as

$$Q_{fuel} = \frac{q_i'' A_{fuel} \Delta H_c}{L_v} \tag{3.21}$$

where q_i'' is the incident radiant heat flux reaching the fuel surface (MW/m²), and its value is generally around 70 kW/m² in post-flashover fire, A_{fuel} is the exposed fuel surface area (m²), ΔH_c is the heat of combustion of volatiles (MJ/kg), and L_v is the heat of gasification (MJ/kg). The amount of energy required to pyrolyse a unit fuel mass is called heat of gasification. The value of heat of gasification for wood ranges between 1.7 and 5.9 MJ/kg and for plastic ranges between 1.2 and 3.7 MJ/kg (Drysdale, 1998).

The underlying assumption in Eq. (3.21) is that the HRR Q_{fuel} is proportional to the incident heat flux without any influence of the shape and thickness on the burning of fuel. It is appropriate for liquid or plastic fuels. For fuels like wood, on which the char layer forms on burning, the burning rate also depends on the thickness and size of the fuel. The average HRR Q_{fuel} (MW) for such fuels is given as

$$Q_{fuel} = v_p \rho A_{fuel} \Delta H_c \qquad (3.22)$$

where ρ is the density of fuel (kg/m³), and v_p is the surface regression rate of the fuel (m/s). Babrauskas (1981) reported the surface regression rate for burning thick wood slabs as 8.5–10×10^{-6} m/s (0.5–0.6 mm/min). For thin wood slabs, the regression rate increases as a function of thickness as $2.2 \times 10^{-6}\, D^{-0.6}$ m/s, where D is the thickness of slab (m).

In any of the above cases, by considering the constant HRR Q_{fuel}, the duration of burning period t_b (s) can be calculated as

$$t_b = \frac{E}{Q_{fuel}} \qquad (3.23)$$

where E is the total amount of fuel in terms of energy in the compartment (MJ). In worked examples, it is observed different answers of HRR Q_{fuel} for different equations for fuel-controlled fire, which suggests more research is required in this area. It is of not much importance because most post-flashover fire calculations assume ventilation control burning.

3.14 TIME-TEMPERATURE CURVES

The performance of structural members under a given fire scenario depends upon the fire severity expressed using time-temperature curves to which the structure is exposed. Structural analysis for fire resistance is done for a given design fire (time-temperature curve) exposure. Various time-temperature curves are discussed in this section.

3.14.1 STANDARD FIRE

Building codes of most countries in the world specify the fire resistance of the structural members and assemblies in terms of fire ratings depending on the type and occupancy of the building. To achieve uniformity in fire ratings established by various countries worldwide, fire resistance ratings for members or assemblies are established through fire test standards using standard time-temperature curves. Standard fire exposure is a widely used fire scenario to compare structural fire performance of building materials and structural members. The standard fire scenario assumes that mostly the fire occurs by burning cellulosic material (wood) that grows in a typical compartment and does not include any decay phase. Widely used test specifications are ASTM E119 (ASTM, 2020) and ISO 834 (ISO, 2012), which are compared in this section. Other international fire resistance test standards specify similar time-temperature curves (Lie, 1995).

The ISO 834 (ISO, 2012) time-temperature relation is given as

$$T = 345\log(8t + 1) + T_0 \tag{3.24}$$

where t is time (min), and T_0 is the ambient temperature (°C).

Lie (1995) gave several equations for approximating the curve given in the form of discrete points in ASTM E119, the simplest of which evaluates the temperature (°C) as

$$T = 750\left[1 - e^{-3.79553\sqrt{t_h}}\right] + 170.40\sqrt{t_h} + T_0 \tag{3.25}$$

where t_h is the time (h), and T_0 is the ambient temperature (°C).

Eurocode 1 (CEN, 2002) proposed a separate equation representing the fire curve for external fires. External fire scenario is used for the structural design of exterior members in case of fire burning outside the compartment/separating walls and is represented by the following equation:

$$T = 660\left[1 - 0.687e^{-0.32t} - 0.313e^{-3.8t}\right] + T_0 \tag{3.26}$$

where t is the time (min), and T_0 is the ambient temperature (°C).

Fire resulting from burning petrochemicals and plastics is much more severe and spreads rapidly than the standard fire curve for cellulosic materials. A separate fire scenario (time-temperature curve) defined as a standard hydrocarbon fire curve is adopted. The hydrocarbon fire curve relation given by ASTM E1529 (ASTM, 2022) is

$$T = 1,080\left[1 - 0.325e^{-0.167t} - 0.675e^{-2.5t}\right] + T_0 \tag{3.27}$$

where t is the time (min), and T_0 is the ambient temperature (°C).

All the standard fire curves, the external fire curve, and the standard hydrocarbon fire curve given in ASTM E1529 and Eurocode are compared in Figure 3.17.

3.15   DESIGN FIRES

Standard fire scenarios discussed in the previous sections do not represent the real fire scenarios in buildings because of the underlying assumptions, such as the absence of decay phase in standard fires, which is present in a real fire scenario. Hence, to understand the structural behavior in realistic fire, it is necessary to arrive at real fire scenarios that account for fuel load, ventilation conditions, and compartment characteristics. Fire resistance ratings in codes published using standard fire exposure (from standard test) are related to the expected real fire scenarios using the concept of equivalent fire severity. Fire design standards of different countries provide simplified relations for calculating temperature-time based on fire load and ventilation conditions. These relations are classified as hand calculations, published curves, parametric fire equations, and computer models based on the complexity level.

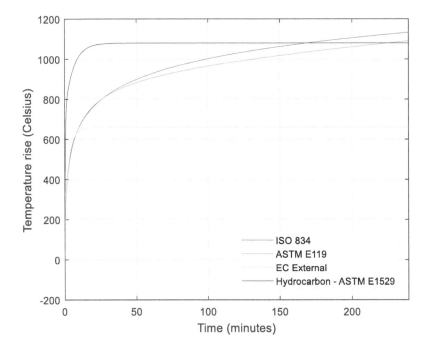

FIGURE 3.17 Comparison of different standard fire curves, hydrocarbon curve, and external fire curve.

3.15.1 CONSTANT PEAK TEMPERATURE

A straightforward method is to assume a constant temperature throughout the burning period to evaluate fire resistance. Such a time-temperature curve is sufficiently accurate for simple designs. The duration of the burning period can be obtained as discussed earlier, assuming ventilation-controlled fire.

This approach does not distinguish between different stages of fire development and their corresponding temperature and only uses a single peak temperature reached during the burning stage. Peak temperatures are generally observed in the post-flashover fire when all the items are burning, and their estimation is essential for structural design for fire safety. Several experimental studies have shown scatter results for temperature in post-flashover fires.

A large number of wood crib fires in small-scale compartments were reported by Thomas and Heselden (1972). The recorded temperature was the average of several thermocouple readings within each compartment. The maximum recorded temperatures during steady phase burning in the experiments are plotted as a function of the ventilation factor in Figure 3.18. Law (1983) developed an empirical relation for the line in Figure 3.18 for the maximum temperature T_{max} (°C) as

$$T_{max} = \frac{6,000\left(1 - e^{-0.1\Omega}\right)}{\sqrt{\Omega}} \tag{3.28}$$

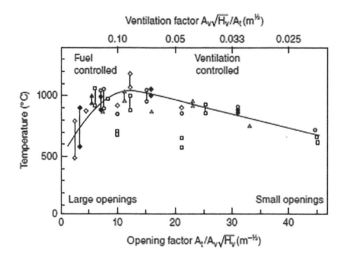

FIGURE 3.18   Maximum temperatures in experimental fires (Thomas and Heselden, 1972).

For small fuel loads, the maximum temperature in Eq. (3.28) may not be reached; therefore, it can be reduced as

$$T = T_{max}\left(1 - e^{-0.05\psi}\right) \tag{3.29}$$

where $\psi = \dfrac{L}{\sqrt{A_v(A_t - A_v)}}$ where L is the fire load (kg, wood equivalent).

3.15.1.1   Published Curves

Through experiments, Magnusson and Thelandersson (1970) investigated the effects of fuel load and ventilation conditions on fire growth characteristics. They published the time-temperature curves, often referred to as 'Swedish' fire curves, by applying heat balance principles. The curves were created for different ventilation factors and varying fuel loads for certain compartment geometries. It was found that for a constant fuel load, well-ventilated fires burnt faster and attained higher temperatures in a short span of time as opposed to poorly ventilated fires that burnt slowly and attained lower temperatures over longer periods of time.

3.16   PARAMETRIC CURVES

Eurocode 1 (CEN, 2002) gives time-temperature relation referred to as parametric fire to represent the compartment fires by accounting for fuel load, ventilation openings, and lining materials. The time-temperature relation given in Eurocode 1 (CEN, 2002) is

$$T = 1,325\left(1 - 0.324e^{-0.2t^*} - 0.204e^{-1.7t^*} - 0.472e^{-19t^*}\right) \tag{3.30}$$

where T is the temperature (°C), and $t^* = \Gamma t$ is fictitious time (h), where t is the actual time (h), and Γ, which governs the temperature rise rate, is given by

$$\Gamma = \frac{\left(F_v/F_{ref}\right)^2}{\left(b_v/b_{ref}\right)^2} \tag{3.31}$$

where $b = \sqrt{k\rho C_p}\left(\frac{Ws^{0.5}}{m^2K}\right)$, is the thermal inertia; b_{ref} is the reference value of thermal inertia, $F_v = A_v\sqrt{H_v}/A_t\left(m^{0.5}\right)$ is the ventilation factor, and F_{ref} is the reference ventilation factor.

The parametric fire equation (3.30) is a good approximation of the standard fire curve for temperatures up to 1,300°C. Hence, for the special case of $F_v = F_{ref}$ and $b_v = b_{ref}$, Eurocode parametric fire curve is close to ISO 834 standard fire curve. Eurocode 1 (CEN, 2002) gave the values of F_{ref} and b_{ref} as 0.04 and 1,160, respectively, based on the standard fire ISO 834 in a compartment. For these values, the equation of Γ is

$$\Gamma_{EC} = \frac{\left(F_v/0.04\right)^2}{\left(b_v/1,160\right)^2} \tag{3.32}$$

Studies by Feasey and Buchanan (2002) showed that the temperatures predicted by the above value of b_{ref} are often too low and revised the value of b_{ref} as 1,900 for better estimation of temperatures using Eq. (3.30):

$$\Gamma_{Design} = \frac{\left(F_v/0.04\right)^2}{\left(b_v/1,900\right)^2} \tag{3.33}$$

The parametric fire curve discussed above assumes the boundaries of the compartment (walls and ceiling) comprise of single layer of material. For two or more layers of materials, Eurocode 1 (CEN, 2002) provides a sequential method to calculate the effective value of the b term.

3.17 BURNING PERIOD

Eurocode 1 also simplifies the equation for the duration of burning period t_d (h) for a ventilation-controlled fire in a compartment and is given as

$$t_d = \frac{0.0002q_{t,d}}{F_v} \tag{3.34}$$

where $q_{t,t}$ is the design FLED (MJ/m² of the total surface area A_t).

For fuel-controlled fires, limioting burning duration (t_{lim}) is 25, 20 and 15 minutes, depending upon the slow, medium and fast fire growth rate, respectively.

3.17.1 DECAY RATE

Eurocode 1 (2002) also gave the time-temperature relationship for the decay (cooling) phase of the parametric fire as

$$T = T_{max} - 625\left(t^* - t_{max}^* x\right) \text{ for } t_{max}^* \leq 0.5$$

$$T = T_{max} - 250\left(3 - t_{max}^*\right)\left(t^* - t_{max}^* x\right) \text{ for } 0.5 < t_{max}^* < 2.0$$

$$T = T_{max} - 250\left(t^* - t_{max}^* x\right) \text{ for } t_{max}^* \geq 2.0 \tag{3.35}$$

where $t_{max}^* = \left(\dfrac{0.0002q_{t,d}}{F_v}\right)\Gamma$ and $x = \{1.0 \text{ if } t_{max} > t_{lim} \quad \dfrac{t_{lim}\Gamma}{t_{max}^*} \text{ if } t_{max} = t_{lim}$ and

$t_{lim} = 25$, 20 or 15 min.

3.18 TIME-TEMPERATURE CURVES

Eurocode time-temperature relations discussed above are plotted in Figure 3.19 for various ventilation factors, fuel loads, and materials. These plots were developed based on actual fire load surveys conducted by Khan and Srivastava (2018) in Gujarat, India.

A Matlab function to calculate the parametric fire curve given the FLED, thermal inertia, ventilation factor, and growth rate of fire (that governs t_{lim}) is given in `ec _ parametric _ curve.m`.

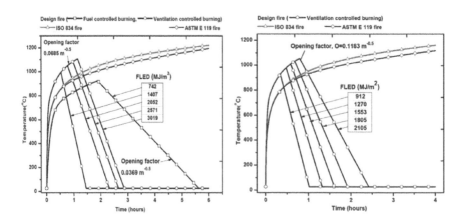

FIGURE 3.19 Eurocode parametric fire curves under different conditions based on field surveys (Khan and Srivastava, 2018).

ec_parametric_curve.m

```
function [ts, T] = ec_parametric_curve(qtd, Fv, bv, tlim)
% Input parameters
%qtd     = design FLED, MJ/m^2 (w.r.t. total internal area)
%Fv      = Ventilation factor, m^0.5
%bv      = thermal inertia, Ws^0.5/m^2K
%tlim    = minutes
% tlim = 25 for slow, 20 for medium, and 15 for fast growth
rate of fire
%
% ts: time vector in minutes
% T : parametric temperatures calculated as per Eurocode
 % Do not change below this.
if Fv<0.02 || Fv>0.2
   warning('Ventilation factor not in the range [0.02,
0.20]');
end
if qtd<50 || qtd>1,000
   warning('FLED not in the range [50, 1,000]');
end
if bv<100 || bv>2,200
   warning('Thermal inertia not in the range [100, 2,200]');
end
 % Reference parameters
Fv_ref  = 0.04;
bv_ref  = 1,160;
% Gamma factor
gamma  = ((Fv/bv)/(Fv_ref/bv_ref))^2;
 % limiting values
Fv_lim = 0.1e-3 * qtd / tlim;
gamma_lim = (Fv_lim/bv)^2/(Fv_ref/bv_ref);
if Fv>0.04 && qtd<75 && b<1,160
   kg = 1+(Fv/Fv_ref - 1)*(qtd/75 - 1)*(1 - bv/bv_ref);
   gamma_lim = kg * gamma_lim;
else
   gamma_lim = gamma;
end
 % Time up to which temperature will rise
tlim = tlim/60;
tmax = round(max(0.0002*qtd / Fv, tlim), 4);
tmax_star = gamma_lim * tmax;
if tmax == tlim
   disp('Fire is fuel controlled');
else
   disp('Fire is ventilation controlled');
end
% times at which temperature is calculated (min)
ts = (0:0.1:180)/60;
ts_star = gamma * ts;
```

```
% heating phase temperature (Celsius)
T = 20+1,325 * (1 - 0.324*exp(-0.2*ts_star) - ...
                    0.204*exp(-1.7*ts_star) - ...
                    0.472*exp(-19*ts_star));

% calculate cooling phase
x = 1;
if tmax == tlim
    x = tlim * gamma / tmax_star;
end
max_idx = find(ts_star > tmax_star);
Tmax = T(max_idx(1)-1);
 if tmax_star <= 0.5
    T(max_idx) = Tmax - 625*(ts_star(max_idx) - ts_max*x);
elseif tmax_star < 2
    T(max_idx) = Tmax - 250*(3 - tmax_star)*(ts_star(max_idx)
- tmax_star*x);
else
    T(max_idx) = Tmax - 250*(ts_star(max_idx) - tmax_star*x);
end
T = max(T, 0);
 %plot(ts*60, T); box on; grid on;
%xlabel('Time (minutes)'); ylabel('Temperature (Celsius)');
```

3.19   FIRE RESISTANCE

Fire resistance of building components is defined in terms of fire rating of their functional aspects. These include

- Strength rating: strength fire rating refers to the time up to which a structural member can fulfil the load demand when subjected to a standard fire. This is typically applicable to all load bearing members of a building such as beams, columns, slabs, and load-bearing walls.
- Insulation rating: insulation fire rating refers to the time up to which a building component can maintain the rise of temperature on its unexposed side within a threshold temperature. The threshold is typically defined as: average temperature rise up to 140°C, maximum temperature rise up to 180°C and maximum actual temperature up to 220°C. This rating is typically applicable to partitioning elements such as partition walls, fire doors, and slabs. Materials with good insulation rating ensure that combustibles on the other side of the partition will not ignite due to radiation (caused by temperature of the unexposed side of the partition element).
- Integrity rating: integrity fire rating refers to the time up to which a building component resists formation of fissures or cracks that can allow the passage of hot gases and fire to the other side of the compartment and is applicable

to partition elements, as in insulation rating. The measurements are usually carried out by placing a cotton pad of specified dimensions and thickness over the visually observed cracks; an ignition of the cotton pad indicates failure of integrity of the building element.
- Smoke rating: smoke rating refers to the time up to which a building component does not allow passage of smoke beyond a certain threshold. This rating is applicable typically to firestop systems that are used to seal structural or service gaps of a building.

The aforementioned fire resistance ratings indicate two important aspects: first, fire ratings are measured in the units of time (typically in h or min), and second, they are always measured with respect to a standard fire curve (e.g., ISO 834 or ASTM E119). It can be observed that an assessment of fire ratings requires calculation of strength of a member (in case strength rating is needed) and temperature (for all ratings). Certain codes offer simplified assessment of strength rating through temperature stipulations directly and in such cases, strength ratings can be calculated purely based on thermal analysis.

It is important to correlate these standard fire ratings to real design scenarios, i.e., how much fire rating should one choose when designing a building. A correspondence between the required standard fire rating based on the possible fire scenarios of a building is based on the concept of equivalent fire severity, which is discussed next.

3.20   EQUIVALENT FIRE SEVERITY

Equivalence between the standard fire and a real fire is usually defined via three metrics which are discussed here. First, one would generate the possible real fire curves (time-temperature) for a given building or compartment depending on its occupancy, expected fire load energy density, geometry, materials of construction, and ventilation conditions. Once a set of possible real fire curves have been developed (usually through the parametric fire curve discussed earlier), the notion of equivalent fire severity is utilized to back-calculate the required standard fire rating so that the building satisfies its stipulated functional requirements and limits states in the chosen critical fire scenario.

3.20.1   Area Equivalence

Time of exposure in a standard fire is considered equivalent to that of a real fire if the area under the time-temperature curves for both cases is the same. Figure 3.20 illustrates this concept. If A_r is the total area of the temperature-time curve of the real fire from the equivalent exposure time in a standard fire and is defined as the time up to which the area under the standard fire curve A_s is same as that of the real fire. This equivalence concept is the oldest but does not have a sound theoretical basis behind it. However, its use is relatively simple and is not dependent on thermal or structural analysis of the system.

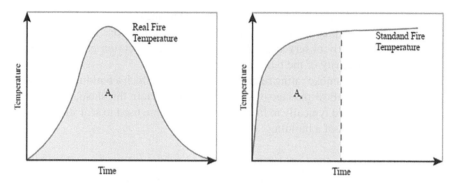

FIGURE 3.20   Area equivalence of real and standard fire curves.

3.20.2   MAXIMUM TEMPERATURE EQUIVALENCE

Maximum temperature concept requires thermal analysis of the system and the equivalent severity of the standard fire is accomplished by computing the time it takes the standard fire to heat a structural member to a temperature that it equals to the maximum temperature the member is expected to reach in a real fire scenario. Figure 3.21 illustrates this concept with respect to a steel structural member. Though it requires thermal analysis, it is a more rational equivalence. Also, it can be observed that for different members (within the same fire scenario), the temperature equivalence (and hence, the required fire rating) can be different.

3.20.3   MINIMUM LOAD CAPACITY EQUIVALENCE

This is similar to the maximum temperature equivalence except that it is applicable only to load-bearing structural members (applicable to strength fire rating) and requires a strength analysis in addition to the thermal analysis. Equivalent severity is achieved when the standard fire degrades the strength of the member to an extent that it is equal to the minimum strength experienced by the structural member during the real-fire scenario. Figure 3.22 illustrates this concept.

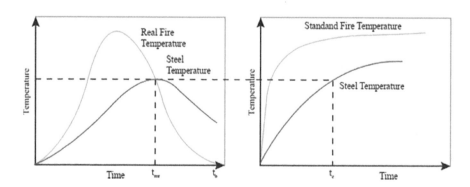

FIGURE 3.21   Temperature equivalence of real and standard fire curves.

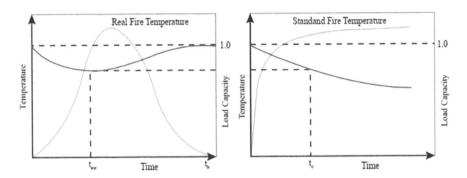

FIGURE 3.22 Strength equivalence of real and standard fire curves.

3.21 THERMODYNAMIC PROPERTIES

This section introduces certain fundamental aspects of heat transfer to enable the reader to gain better understanding of such calculations in the context of structural fire engineering. Discussions will be presented in one-dimensional context for ease of illustration and understanding. Readers interested in more in-depth understanding are encouraged to consult the available textbooks in the area (Bergman et al., 2011; Wickstrom, 2016; Drysdale, 1998). Thermodynamic properties of a system are usually classified into intensive and extensive properties. Intensive or bulk properties do not change with respect to the amount of material. Temperature and density are examples of intensive properties. Extensive properties depend on the amount of material, such as heat and mass. Table 3.3 summarizes some relevant intensive and extensive properties.

A good understanding of the modes of heat transfer is essential in structural fire engineering. All three modes of heat transfer – conduction, convection, and radiation – play an important role in structural heat transfer calculations and will be briefly reviewed in subsequent sections.

3.21.1 CONDUCTION

Conduction is a dominant mode of heat transfer in solids and occurs due to energy transfer from more energetic to less energetic particles (atoms or molecules) of a material. In liquids and gases, conduction does take place but the other modes are more prominent. Heat conduction is governed by the Fourier's law, which states that the rate of heat transfer through a material is proportional to the negative gradient in

---

TABLE 3.3
Intensive and Extensive Properties

Intensive Properties	Extensive Properties
Do not depend on the amount of material	Depend on the amount of material
Temperature, density	Heat, mass

the temperature and to the area perpendicular to the gradient. In a 1D setting, it may be visualized through Figure 3.23 and can be written as

$$\dot{q}_x \propto -A\frac{\mathrm{d}T}{\mathrm{d}x} \tag{3.36}$$

where $\dot{q}_x = \mathrm{d}q_x/\mathrm{d}t$ is the rate of heat transfer in x direction, A is the area perpendicular to x direction, and $\mathrm{d}T/\mathrm{d}x$ denotes the gradient of temperature, T. The negative sign indicates that heat flows from the higher to the lower temperature (i.e., $T_1 > T_2$). This is usually written in a per unit area format as

$$q_x'' \propto -\frac{\mathrm{d}T}{\mathrm{d}x} = -k\frac{\mathrm{d}T}{\mathrm{d}x} \tag{3.37}$$

where $q_x'' = \dot{q}_x/A$ is the heat flux (W), and k is the constant of proportionality, called the thermal conductivity of the material. Fourier carried out a number of experiments on metal to arrive at this relationship; the values of k for different materials are well documented and available through different resources. Through dimensional analysis, one can work out the units of k to be $\mathrm{W}/(\mathrm{m}^2\mathrm{K})$ or $\mathrm{W}/(\mathrm{m}^2{}^\circ\mathrm{C})$.

The schematic shown in Figure 3.23 can be visualized as that of the cross section of a wall (say, of masonry), where the outside temperature is T_1 (at $x = 0$), and the inside temperature is T_2 (at $x = L$.) with L being the thickness of the wall. In such a scenario, the heat flux through the wall under steady-state conditions can be calculated from Eq. (3.37) as

$$q_{cond}'' = q_x'' = -k\frac{\mathrm{d}T}{\mathrm{d}x} = \frac{k}{L}(T_1 - T_2) \tag{3.38}$$

It can be observed that the heat flux comes out to be constant (k is assumed to be independent of temperature here; if k is temperature-dependent, the relation in

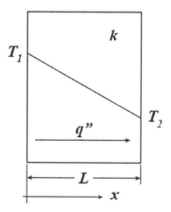

FIGURE 3.23   1D heat conduction.

Eq. (3.38) will need to be modified). Once the heat flux has been calculated, temperature at any point within the wall, T_x, can be calculated as

$$T_x = T_1 - q_x'' \frac{x}{k} \tag{3.39}$$

For the calculation of steady-state temperatures, it is convenient to employ the 'electric circuit analogy', where the temperature difference $\Delta T = T_1 - T_2$ is viewed as voltage difference, heat flux q_x'' is viewed as current, and 'thermal resistance' $R_{cond} = L/k$ is viewed as the resistance. The heat conduction relation can be written as

$$\Delta T = q_{cond}'' R_{cond} \tag{3.40}$$

The principal advantage of such an analogy lies in the fact that when there are multiple materials (e.g., in a composite wall or a fire-protected steel section), the heat flux and temperatures at intermediate materials can be computed conveniently. Further, the 'total' thermal resistance of a composite system can be computed by adding the individual thermal resistances (similar to electrical resistors in series). For a wall comprising of n layers with the leftmost edge at temperature T_0 and the rightmost edge at temperature T_n, as shown in Figure 3.24, the total thermal resistance of the wall can be computed as

$$R_{total} = \sum_{i=1}^{n} R_i \tag{3.41}$$

The total heat flux passing through the wall can be calculated from Eq. (3.40) as

$$q_x'' = \frac{\Delta T}{R_{total}} = \frac{T_0 - T_n}{R_{total}} \tag{3.42}$$

The temperature at the ith material interface can be calculated from Eq. (3.39) as

$$T_i = T_0 - q_x'' \sum_{j=1}^{i} R_j \tag{3.43}$$

From Eq. (3.42), Eq. (3.43) can be rewritten as

$$T_i = T_0 + \frac{\sum_{j=1}^{i} R_j}{R_{total}} (T_n - T_0) \tag{3.44}$$

	k_1	k_2	k_3		k_{n-2}	k_{n-1}	k_n	
T_0	1	2	3	...	$n-2$	$n-1$	n	T_n
	L_1	L_2	L_3		L_{n-2}	L_{n-1}	L_n	

FIGURE 3.24 Wall with n layers.

Example 3.1

A wall consists of 230 mm clay brick masonry, 15 mm mortar plaster, and 12 mm ornamental wooden board with conductivities equal to 0.15, 0.71, and 0.14 W/(m²K), respectively. The outer surface has a constant temperature of 500 °C due to an external fire, and the inside of the wall surface is at a temperature of 30 °C. Calculate the temperatures at the intermediate interfaces (T1 and T2).

Given Data:

Thicknesses:

$t_1 = 0.23\,\text{m}$

$t_2 = 0.015\,\text{m}$

$t_3 = 0.012\,\text{m}$

Thermal conductivities:

$$k_1 = \frac{0.15\,\text{W}}{\text{m}^2 \cdot \text{C}} = 0.15\,\text{W/m}^2 \cdot \text{C}$$

$$k_2 = \frac{0.71\,\text{W}}{\text{m}^2 \cdot \text{C}} = 0.71\,\text{W/m}^2 \cdot \text{C}$$

$$k_3 = \frac{0.14\,\text{W}}{\text{m}^2 \cdot \text{C}} = 0.14\,\text{W/m}^2 \cdot \text{C}$$

Temperatures

$$T_o = 500\,\text{C}$$

$$T_i = 30\,\text{C}$$

Calculate thermal resistances

$$R_1 = \frac{t_1}{k_1} = \frac{0.23\,\text{m}}{0.15\,\text{W/m}^2 \cdot \text{C}} = 1.53\,\text{kg}^{-1} \cdot \text{m} \cdot \text{s}^4 \cdot \text{A}$$

$$R_2 = \frac{t_2}{k_2} = \frac{0.015\,\text{m}}{0.71\,\text{W/m}^2 \cdot \text{C}} = 0.0211\,\text{kg}^{-1} \cdot \text{m} \cdot \text{s}^4 \cdot \text{A}$$

$$R_3 = \frac{t_3}{k_3} = \frac{0.012\,\text{m}}{0.14\,\text{W/m}^2 \cdot \text{C}} = 0.0857\,\text{kg}^{-1} \cdot \text{m} \cdot \text{s}^4 \cdot \text{A}$$

$$R_{tot} = R_1 + R_2 + R_3 = 1.53\,\text{kg}^{-1} \cdot \text{m} \cdot \text{s}^4 \cdot \text{A} + 0.0211\,\text{kg}^{-1} \cdot \text{m} \cdot \text{s}^4 \cdot \text{A} + 0.0857\,\text{kg}^{-1} \cdot \text{m} \cdot \text{s}^4 \cdot \text{A}$$

$$= 1.64\,\text{kg}^{-1} \cdot \text{m} \cdot \text{s}^4 \cdot \text{A}$$

Calculate interface temperatures

$$T_1 = T_o + \frac{R_1}{R_{tot}}(T_i - T_o) = 500°\text{C} + \frac{1.53\,\text{kg}^{-1} \cdot \text{m} \cdot \text{s}^4 \cdot \text{A}}{1.64\,\text{kg}^{-1} \cdot \text{m} \cdot \text{s}^4 \cdot \text{A}}(30\text{C} - 500°\text{C}) = 60.62°\text{C}$$

3.21.2 CONVECTION

Convection is said to take place when energy transfer occurs through bulk motion of a fluid (liquid or gas). It is one of the modes through which heat transfer occurs from the source of fire to a structural member during a fire. The rate of convective heat transfer is given by Newton's law of cooling as

$$q''_{conv} = h_c \left(T_\infty - T_s \right) \tag{3.45}$$

where T_s and T_∞ are the surface and the fluid temperatures, respectively, and h_c is the convective heat transfer coefficient. h_c has the units $W/(m^2K)$ and is influenced by the surface geometry, the nature of fluid motion, and certain thermodynamic and transport properties of the fluid. h has been determined experimentally for a number of scenarios. Here, the convective heat flux is positive when the heat transfer takes place from the surroundings to the surface (i.e., $T_s < T_\infty$ implying that the surroundings are hotter than the surface – typically the case during a fire). Convection can be either forced (e.g., through a fan) or natural.

Similar to conduction, the (surface) thermal resistance, R_h, can be defined as

$$R_{conv} = \frac{1}{h_c} \tag{3.46}$$

yielding the convective heat flux to be $q''_{conv} = \Delta T / R_{conv}$ with $\Delta T = (T_\infty - T_s)$. The use of convective thermal resistance for steady-state calculations is illustrated through an example.

Example 3.2

A masonry wall of 230 mm thickness is exposed to gas at Tg = 400 °C on one side, while the ambient temperature on the other side of the wall is fixed at T2 = 30 °C. Calculate the surface temperature T1 of the wall on the exposed side. Assume the conductivity of the masonry to be k = 0.15 W/mK and convective heat transfer coefficient h = 5 W/(m²K).

Given data

$$h = \frac{5\,W}{m^2 \cdot C} = 5\,W/m^2 \cdot C$$

$$k = \frac{0.15\,W}{m \cdot C} = 0.15\,W/m \cdot C$$

$$L = 0.23\,m$$

$$T_g = 400\,°C$$

$$T_2 = 30\,°C$$

Calculate thermal resistances

$$R_h = \frac{1}{h} = \frac{1}{5\,\text{W/m}^2 \cdot \text{C}} = 0.2\,\text{m}^2 \cdot \text{C/W}$$

$$R_k = \frac{L}{k} = \frac{0.23\,\text{m}}{0.15\,\text{W/m} \cdot \text{C}} = 1.53\,\text{m}^2 \cdot \text{C/W}$$

$$R_{tot} = R_h + R_k = 0.2 + 1.53 = 1.73\,\text{m}^2 \cdot \text{C/W}$$

Calculate surface temperature

$$T_1 = \frac{R_h \cdot T_2 + R_k \cdot T_g}{R_{tot}} = \frac{0.2\,\text{m}^2 \cdot \text{C/W} \cdot 30\,\text{C} + 1.53\,\text{m}^2 \cdot \text{C/W} \cdot 400\,°\text{C}}{1.73\,\text{m}^2 \cdot \text{C/W}} = 357.31\,°\text{C}$$

3.21.3 RADIATION

Radiation involves energy transfer through electromagnetic waves towards the infra-red portion of the spectrum (wavelength ranging from 0.4 to $100\mu m$). Thermal radi-ation is emitted by all bodies that are at non-zero (absolute) temperature. Unlike conduction and convection, radiation does not require the presence of a medium for heat transfer. The notion of a black body is often used in discussions pertaining to radiation. An ideal body which is a perfect emitter or absorber of energy is called a black body. The radiative heat flux emitted by a black body at a temperature T_s (K) is given by the Stefan–Boltzmann law as

$$q_e'' = \sigma T_s^4 \tag{3.47}$$

where $\sigma = 5.670 \times 10^{-8}\,\text{W/}\left(\text{m}^2\text{K}\right)$ is the Stefan–Boltzmann constant. It should be noted that in radiation-related calculations, temperatures must be used in the units of Kelvin. Furthermore, Eq. (3.47) denotes the heat flux going out of the surface of the body. Real materials are mostly considered as non-ideal grey bodies, where Eq. (3.47) denotes the upper limit of the heat flux emitted by the surface. A material parameter, emissivity, is introduced to determine the actual heat flux emitted by a real body as

$$q_e'' = \epsilon_s \sigma T_s^4 \tag{3.48}$$

where $0 < \epsilon_s \leq 1$ (unitless) is the emissivity of the body. In the context of radiative heat transfer, the notions of incidence, absorption, and reflection are central. Imagine staying outside on a sunny day. The radiative heat flux emitted by the Sun can be cal-culated from Eq. (3.47) as the estimates for the surface temperature are known. This heat flux (emitted by the Sun) acts as an incident heat flux on any object that obstructs its path. The object will reflect part of this incident radiation and absorb the remain-ing portion. The object, being at a certain temperature, will have its own emitted heat flux as well. Note that in case of semi-transparent bodies, transmittance should also be accounted for. The overall energy balance can be visualized through Figure 3.25.

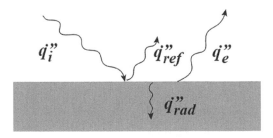

FIGURE 3.25 Energy balance of a body with incident radiation.

The portion of the incident radiation absorbed by the body is calculated as

$$q''_a = \alpha_s q''_i \tag{3.49}$$

where q''_i is the incident radiative heat flux, q''_a is the heat flux absorbed by the body, and α_s is the absorptivity of the surface (a material parameter). Owing to the first law of thermodynamics, through Kirchhoff's identity, $\alpha_s = \epsilon_s$. The remainder of the incident flux is reflected by the body and is calculated as

$$q''_r = q''_i - q''_a = (1 - \epsilon_s) q''_i \tag{3.50}$$

Thus, the net heat flux acting on the body due to radiation becomes

$$q''_{rad} = q''_a - q''_e = \epsilon_s \left(q''_i - \sigma T_s^4 \right) \tag{3.51}$$

Using Eq. (3.51), one can calculate the net radiative heat transfer from one body (at temperature T_1 with emissivity ϵ_1) to a second body (at temperature T_2 with emissivity ϵ_2) as

$$q''_{rad,1-2} = \sigma \epsilon_2 \left(\epsilon_1 T_1^4 - T_2^4 \right) \tag{3.52}$$

During a fire scenario, one can have multiple emitting surfaces leading to multiple sources of incident heat flux. It is usual to enclose emitting surfaces in imaginary boxes for ease of analysis. If two infinite plates are kept parallel to each other, the net radiative heat flux from one plate to the other is given by

$$q''_{rad,1-2} = \sigma \epsilon_{res} \left(T_1^4 - T_2^4 \right) \tag{3.53}$$

where $\epsilon_{res} = \dfrac{1}{\dfrac{1}{\epsilon_1} + \dfrac{1}{\epsilon_2} - 1}$ is the resultant emissivity. Furthermore, in certain scenarios,

one may assume that one of the emitting surfaces (typically fire) is a perfect black body: $\epsilon_1 = 1$, which leads to the following condition:

$$q''_{rad,1-2} = \sigma \epsilon_2 \left(T_1^4 - T_2^4 \right).$$

Through Eq. (3.53), one may define a radiative heat transfer coefficient h_r as

$$h_r = \sigma \epsilon_{res} \left(T_1 + T_2 \right) \left(T_1^2 + T_2^2 \right) \tag{3.54}$$

which allows defining the radiative heat flux in a manner similar to that of convective or conductive heat flux as

$$q''_{rad,1-2} = h_r \left(T_1 - T_2 \right) = h_r \Delta T \tag{3.55}$$

One may now define a radiative thermal resistance as

$$R_{rad} = \frac{1}{h_r} \tag{3.56}$$

However, the use of radiative thermal resistance for heat transfer calculations is not straightforward due to its dependence on temperatures (essentially, it is a non-linear thermal resistance).

Two important laws that govern radiation are the inverse square law and the Lambert's cosine law. Radiation follows Huygen's principle of radial expansion as shown in Figure 3.26. Given a source of heat $\dot{q}$ (typically measured in Watts) at a point, the intensity of heat radiation, I is a measure of its distribution per unit area and per solid angle in a particular direction. Thus, the heat intensity at a distance r over the entire surface of an imaginary sphere shown in Figure 3.26 will be

$$I = \frac{\dot{q}}{4\pi r^2} \tag{3.57}$$

Here, an integration over the entire surface of the sphere of radius r has been considered to obtain the total intensity at a distance r from the source. The units of I will thus be $W/\left(m^2 \cdot sr \right)$.

When dealing with a surface source, typical in fire scenarios, the direction of irradiation from the surface also affects the intensity through Lambert's cosine law (see Figure 3.27) as

$$I_\theta = I \cos\theta \tag{3.58}$$

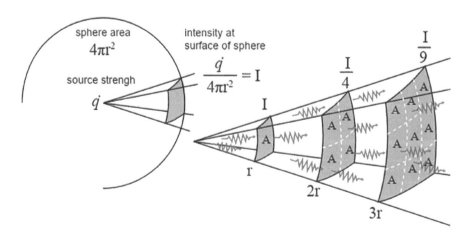

FIGURE 3.26 Radiation.

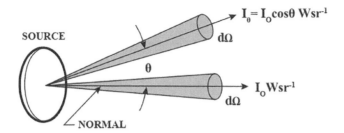

FIGURE 3.27 Lambert's cosine law.

Here, θ is the angle measured form the normal to the surface at which the intensity is being measured.

In fire scenarios, it is usual to have interaction of two surfaces. For instance, a wall being heated from fire from one side will radiate on the other side. Many times, a fire itself is idealized to be enclosed in an imaginary cuboid whose surfaces are assumed to radiate the heat generated by the fire. Thus, consideration of two radiating surfaces arbitrarily oriented with respect to each other is necessary, as shown in Figure 3.28.

Let the areas of the surfaces be A_1 and A_2 and the power of source at surface 1 be $\dot{q}_1$. The radiative heat intercepted by surface 2 will be

$$\dot{q}_{1\to2} = F_{1\to2}\dot{q}_1 \tag{3.59}$$

and the intercepted heat flux will be

$$q''_{1\to2} = F_{1\to2}A_1q''_1 = F_{1\to2}A_1\sigma\epsilon_1\left(T_1^4 - T_2^4\right) \tag{3.60}$$

Here, $F_{1\to2}$ is the configuration factor or the view factor given by

$$F_{1\to2} = \frac{1}{A_1} \int_{A_1} \int_{A_2} \frac{\cos\theta_1\cos\theta_2}{\pi r^2} dA_1 dA_2 \tag{3.61}$$

Table 3.4 provides view factors for some common configurations.

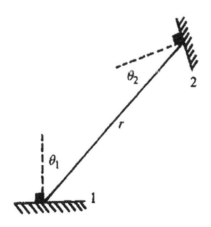

FIGURE 3.28 Effect of orientation on radiative heat flux.

TABLE 3.4
Configuration Factors for Common Geometries

S. No.	Geometry	Configuration Factor
1	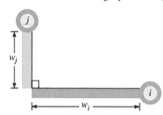 Parallel plates (infinitely long in the third dimension)	$W_i = \dfrac{w_i}{L}, W_j = \dfrac{w_j}{L}$ $F_{ij} = \dfrac{\sqrt{\left(W_i + W_j\right)^2 + 4} - \sqrt{\left(W_i - W_j\right)^2 + 4}}{2 \_ W_i}$
2	Inclined plates (infinitely long in the third dimension, having equal widths)	$F_{ij} = 1 - \sin\left(\dfrac{\alpha}{2}\right)$
3	Plates perpendicular to each other (infinitely long in the third dimension, having unequal widths)	$F_{ij} = \dfrac{1 + \dfrac{w_j}{w_i} - \sqrt{1 + \left(\dfrac{w_j}{w_i}\right)^2}}{2}$
4	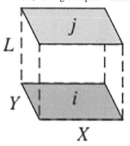 Parallel rectangular plates (finite dimensions)	$X_1 = \dfrac{X}{L}, Y_1 = \dfrac{Y}{L}$ $F_{ij} = \dfrac{2}{\pi X_1 Y_1} X_1 \left\{ \ln\left[\dfrac{\left(1 + X_1^2\right)\left(1 + Y_1^2\right)}{1 + X_1^2 + Y_1^2}\right]^{0.5} \right\}$ $+ X_1\sqrt{1 + Y_1^2}\ \tan^{-1}\dfrac{X_1}{\sqrt{1 + Y_1^2}}$ $+ Y_1\sqrt{1 + X_1^2}\ \tan^{-1}\dfrac{Y_1}{\sqrt{1 + X_1^2}}$ $- X_1 \tan^{-1} X_1 - Y_1 \tan^{-1} Y_1$

(Continued)

TABLE 3.4 *(Continued)*
Configuration Factors for Common Geometries

S. No.	Geometry	Configuration Factor
5	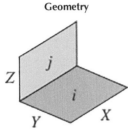 Perpendicular rectangular plates (finite dimensions)	$H = \dfrac{Z}{X},\ W = \dfrac{Y}{X}$ $F_{ij} = \dfrac{1}{\pi W}\left(W\tan^{-1}\dfrac{1}{W} + H\tan^{-1}\dfrac{1}{H} \right.$ $\left. -\sqrt{H^2 + W^2}\ \tan^{-1}\dfrac{1}{\sqrt{H^2 + W^2}} \right.$ $\left. +\dfrac{1}{4}\ln\left\{ \dfrac{(1+W^2)(1+H^2)}{1+W^2+H^2} \times \left[\dfrac{W^2(1+W^2+H^2)}{(1+W^2)(1+H^2)} \right]^{W^2} \times \left[\dfrac{H^2(1+W^2+H^2)}{(1+W^2)(1+H^2)} \right]^{H^2} \right\} \right)$
6	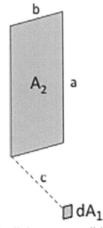 Small radiation source parallel to a large object	$X = \dfrac{a}{c},\ Y = \dfrac{b}{c}$ F_{d1-2} $= \dfrac{1}{2\pi}\left[\dfrac{X}{\sqrt{1+X^2}}\tan^{-1}\dfrac{Y}{\sqrt{1+X^2}} + \dfrac{Y}{\sqrt{1+Y}}\tan^{-1}\dfrac{x}{\sqrt{1+X^2}} \right]$

(Continued)

TABLE 3.4 (*Continued*)

Configuration Factors for Common Geometries

S. No.	Geometry	Configuration Factor
7	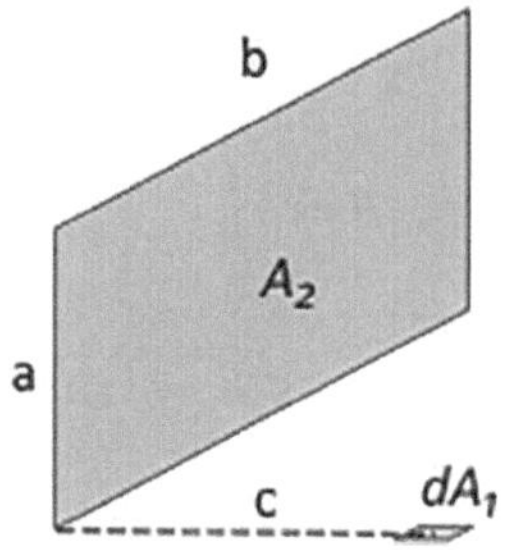 Small radiation source perpendicular to a large object	$F_{d1-2} = \dfrac{1}{2\pi}\left[\tan^{-1}\dfrac{b}{c} - \dfrac{c}{\sqrt{a^2+c^2}}\tan^{-1}\dfrac{b}{\sqrt{a^2+c^2}}\right]$

Example 3.3

An uninsulated fire door made of steel becomes uniformly heated to a temperature of 650°C during a fire. Calculate the maximum incident radiation q″inc, max to a parallel surface 0.5 m from the door leaf with dimensions 1.0 m by 2.1 m and an emissivity of 0.9.

Given data:

$$a = \frac{2.1}{2} = 1.05$$

$$b = \frac{1}{2} = 0.5$$

$$c = 0.5$$

$$\varepsilon = 0.9$$

$$\sigma = \frac{5.67 \cdot 10^{-8}\,\text{W}}{\text{m}^2 \cdot \text{K}^4} = 5.67 \times 10^{-08}\,\text{W/m}^2 \cdot \text{K}^4$$

$$T = 650\,\text{K} + 273\,\text{K} = 923\,\text{K}$$

Calculate view factor

$$X = \frac{a}{c} = \frac{1.05}{0.5} = 2.1$$

$$Y = \frac{b}{c} = \frac{0.5}{0.5} = 1$$

$$F_d = \frac{1}{2 \cdot \pi} \left(\frac{X}{\sqrt{1 + X^2}} \cdot atan \left(\frac{Y}{\sqrt{1 + X^2}} \right) + \frac{Y}{\sqrt{1 + Y^2}} \cdot atan \left(\frac{X}{\sqrt{1 + Y^2}} \right) \right)$$

$$= \frac{1}{2 \cdot 3.14} \left(\frac{2.1}{\sqrt{1 + 2.1^2}} \cdot atan \left(\frac{1}{\sqrt{1 + 2.1^2}} \right) + \frac{1}{\sqrt{1 + 1^2}} \cdot atan \left(\frac{2.1}{\sqrt{1 + 1^2}} \right) \right) = 9.65$$

$$F_t = 4 \cdot F_d = 4 \cdot 9.65 = 38.6$$

Calculate max incident radiation

$$q''_{inc,max} = F_t \cdot \varepsilon \cdot \sigma \cdot T^4 = 38.6 \cdot 0.9 \cdot 5.67 \times 10^{-08} \, \text{W/m}^2 \cdot \text{K}^4 \cdot (923 \, \text{K})^4 = 1,429,611 \, \text{W/m}^2$$

3.22   TRANSIENT HEAT TRANSFER

In most real scenarios, the heat transfer is unsteady or transient. For instance, the incident heat flux on a structure under a real fire or the temperature of a fire will change with time. Consequently, a steady-state assumption may not always be valid. In order to consider the effects of transient heat transfer, another view of heat stored in a material can be considered. Recall that Fourier's law established that the rate of change of heat is proportional to the negative of area and temperature gradient (Eq. 3.36). Converting this relationship to an equation involved introduction of a material parameter, thermal conductivity (k).

One may recall two more fundamental material properties: density (ρ) and specific heat capacity (c). Density is the mass per unit volume of a material (kg/m^3), while specific heat capacity is the quantity of heat absorbed per unit mass of a material when its temperature increases by one unit (J/kg$^\circ$C). This gives another view of writing the total heat of a material as

$$q = \rho c V T \tag{3.62}$$

where V is the total volume of the material. In general, ρ or C may vary with temperature and location within the material. This leads us to the following definition of heat flux

$$\dot{q}''_x - \rho C \frac{dT}{dt} dx \tag{3.63}$$

where the time derivative has operated on T and $V = Adx$ has been considered. Since both Eqs. (3.63) and (3.37) define heat flux, they must be equal, which leads to the transient (1D) heat transfer equation as

$$\rho C \frac{\partial T}{\partial t} = \frac{\partial}{\partial x} \left(k \frac{\partial T}{\partial x} \right) \tag{3.64}$$

If thermal conductivity is constant, a simpler form can be written as

$$\alpha \frac{\partial T}{\partial t} = \frac{\partial^2 T}{\partial x^2} \quad (3.65)$$

where $\alpha = k/(\rho c)$ is called the thermal diffusivity of the material. In the steady-state limit, temperature does not change in time, i.e., $\frac{\partial T}{\partial t} = 0$ as $t \to \infty$.

A usual way of solving the transient heat equation is through the finite difference method, where the derivatives are approximate via differences. Let $T(x,t)$ be the unknown temperature field and $T_i^{(n)} = T(x_i, t_n)$ be the temperature at location x_i and time t_n. The time and spatial derivatives appearing in (3.65) can be approximated using forward and central differencing scheme as

$$\left(\frac{dT}{dt}\right)_i^{(n)} \approx \frac{T_i^{(n+1)} - T_i^{(n)}}{\Delta t}, \text{ and } \left(\frac{d^2 T}{dx^2}\right)_i^{(n)} \approx \frac{T_{i+1}^{(n)} - 2T_i^{(n)} + T_{i-1}^{(n)}}{\Delta x^2} \quad (3.66)$$

where Δt and Δx are the time-step and spatial discretization spacing, respectively. After substituting Eqs. (3.66) in (3.65), one obtains

$$\frac{T_i^{(n+1)} - T_i^{(n)}}{\Delta t} = \frac{T_{i+1}^{(n)} - 2T_i^{(n)} + T_{i-1}^{(n)}}{\alpha \Delta x^2} \quad (3.67)$$

which can be used to calculate $T_i^{(n+1)}$ (temperature field at next time step) as

$$T_i^{(n+1)} = T_i^{(n)} + \Delta t \frac{T_{i+1}^{(n)} - 2T_i^{(n)} + T_{i-1}^{(n)}}{\alpha \Delta x^2} \quad (3.68)$$

Note that this requires the information about the temperature field at the initial stage (i.e., at $n = 0$), which is called the initial condition for this system. The solution written in this manner is first-order accurate in time and second-order accurate in space. The solution also requires appropriate boundary conditions which are discussed next.

3.23   BOUNDARY CONDITIONS

Differential equations such as the steady-state or transient heat equations require appropriate boundary conditions to possess a unique solution. One more aspect to note is that the differential equations, in their current form, are for heat conduction through a material. In case of liquids or gases, a convective term will also get added. Furthermore, it is possible to have an internal heat source or sink (e.g., a fire will be a heat source for a simulation where the room is being considered as the volume of interest). In the current context, however, the discussion is restricted to consider heat transfer within solids (structural members comprising of steel, masonry, concrete, etc.). Thus, the current form of heat conduction equation (steady state or transient) is sufficient.

Boundary conditions can be of three types. The first kind is when temperature is specified at the boundary (called the Dirichlet condition), as shown in Figure 3.29. If the one side of a wall is being maintained at a constant temperature (say through air conditioning), it can be considered to be a Dirichlet condition. Since fires are characterized by their time-temperature curves, there is a tendency to use the fire

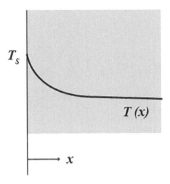

FIGURE 3.29   Dirichlet (temperature) boundary condition.

temperature as a Dirichlet condition on the structural system. However, this can lead to significant overestimation of the structural temperatures and should be avoided. Mathematically, such a condition can be written as $T(x = 0,t) = T_0$, where T_0 is the given temperature; note that in general, T_0 can be time dependent.

The second kind of boundary condition is when heat flux is specified (called the Neumann condition), as shown in Figure 3.30. A special case of the flux condition occurs when the flux is specified to be zero, i.e., no heat can go into or out of the system. This simulates a perfectly insulated condition (sometimes called an adiabatic boundary). It should also be noted that when solving the heat equation using the finite element method, if nothing is specified on a boundary, the boundary condition defaults to a zero flux condition. Since fires can be characterized by HRR vs. time as well, one may consider applying the heat released from the fire as a flux boundary to the structural system. However, this also leads to issues for two reasons: (i) the heat released by a fire is not uniform either in space or in time, and (ii) the heat transfer from the fire to a structure happens through convection and radiation and a simple flux boundary condition fails to capture the necessary physical effects. Mathematically, such a condition can be written as $\dot{q}''\big|_{x_0} = \left(-k\dfrac{dT}{dx}\right)_{x_0} = \dot{q}_0''$, where $\dot{q}_0''$ is the given heat flux at the boundary $x = x_0$.

The third type of boundary condition (called the Robin condition) considers the application of temperature-dependent heat flux at the boundaries and is suitable for

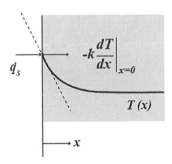

FIGURE 3.30   Neumann (flux) boundary condition.

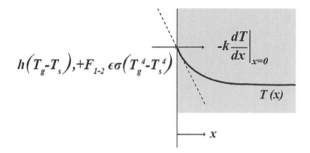

FIGURE 3.31 Robin (mixed) boundary condition.

simulating real fire scenarios, where heat from the fire reaches the structure through convection and radiation, as shown in Figure 3.31. A convective boundary condition can be written as $\dot{q}''|_{x_0} = \left(-k\dfrac{dT}{dx}\right)_{x_0} = h(T_g - T_s)$, where h is the convective heat coefficient, discussed earlier, and T_g is the gas temperature. Note that this also depends on T_s, which needs to be determined as part of the solution. A radiative boundary condition will be written as $\dot{q}''|_{x_0} = F_{1-2}\epsilon\sigma(T_g^4 - T_s^4)$, whereas a combined radiative and convective boundary condition can be written as $\dot{q}''|_{x_0} = h(T_g - T_s) + F_{1-2}\epsilon\sigma(T_g^4 - T_s^4)$. In certain situations, the combined condition can be written as $\dot{q}''|_{x_0} = \bar{h}(T_g - T_s)$, where $\bar{h}$ is a combined convective-radiative coefficient.

3.24 LUMPED CAPACITY IDEALIZATION FOR STEEL

In materials such as steel or aluminum, the rate of conduction is quite fast, and it can be assumed that the temperature of the entire structural member is the same (i.e., conduction is instantaneous). This allows lumping the entire structural member to a single point (zero dimensional) for heat transfer analysis. The overall energy balance simply requires the heat received by the member will convert to stored heat, which in turn raises its temperature. A schematic is shown in Figure 3.32. Since the heat capacity of the entire volume of the material is being lumped to a single point, the name is 'lumped capacity' idealization.

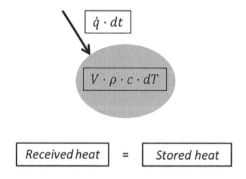

FIGURE 3.32 Energy balance for a lumped body.

Mathematically, the rate of change of temperature can be written as

$$\frac{dT}{dt} = \frac{A}{\rho c V} \dot{q}''$$ (3.69)

where A is the exposed area, and V is the total volume of the structural member. For thin plates, the thickness $d = A/V$ (sometimes also called section factor) can be used if the entire area is exposed. Using the finite difference scheme introduced in Eq. (3.66), the rise in temperature of such a member can be written as

$$\Delta T^{(n)} = T^{(n+1)} - T^{(n)} = \Delta t \frac{A}{\rho c^{(n)} V} \dot{q}''^{(n)}$$ (3.70)

For most structural materials, density is considered to be constant, whereas it is possible to consider the variation of specific heat with temperature in Eq. (3.70). The total heat flux $\dot{q}''^{(n)}$ at any time step can be calculated through the usual energy balance, as the total incident heat flux less than the total heat flux leaving the body (e.g., through radiation). A few common cases are shown in Table 3.5.

Additionally, the time increment to be used in Eq. (3.70) is subjected to a maximum permissible time increment given by

$$\Delta t_{max} = \frac{\rho c V}{h_{total} A}$$ (3.71)

where h_{total} is the total heat transfer coefficient (e.g., for convection and radiation, $h_{total} = h_c + h_r$). If a time increment greater than Δt_{max} is used, the numerical scheme will not converge. However, for good accuracy, it is usually recommended to use time increments in the order of $5\% - 10\%$ of Δt_{max}.

TABLE 3.5

Heat Flux for Different Cases of Lumped Capacity Model

S. No.	Case	$\dot{q}''^{(n)}$
1	Material being heated through convective heat transfer by hot gas	From (45) $\dot{q}''^{(n)} = h_c\left(T_g^{(n)} - T^{(n)}\right)$
2	Material being heated through radiative heat transfer	From (48) $\dot{q}''^{(n)} = \epsilon\sigma T_g^{(n)4}$
3	Material is subjected to a known heat flux	$\dot{q}''^{(n)} = \dot{q}''^{(n)}_{inc}$
4	When material is capable to lose energy through radiation, appropriate reduction from $\dot{q}''^{(n)}$ should be made	$\dot{q}''^{(n)} = \dot{q}''^{(n)} - \epsilon\sigma T^{(n)4}$

Example 3.4

An unprotected aluminum section with a section factor $\dfrac{A}{V} = \dfrac{1}{70}\,m^{-1}$, initially at room temperature of $25°C$, is suddenly exposed to a constant fire temperature of $350°C$. Calculate its temperature as a function of time. Assume aluminium emissivity of 0.9 and a convective heat transfer coefficient of $h_c = 20\ W/(m^2K)$.

The Matlab code for this example is given below. The T-t and q-t graphs are also given.

```
ex_lumped_aluminum.m
% Given data:
d  = 70;        %1/m (= A/V)
Ti = 25+273;    %K (initial temperature)
Tg = 350+273;   %K (exposure temperature)
hc = 20;        %W/(m^2-K)
e  = 0.9;       %emissivity of aluminum
 % Assumed/found data
rho = 2700;     %kg/m^3    (density of aluminum)
c   = 920;      %J/(kg-K)  (specific heat of aluminum)
sbc = 5.67e-8;%W/(m^2-K^4) (Stefan-Boltzmann constant)
dt  = 10;       %s (chosen time increment)
tmax= 3600;     %s (time up to which temperatures are to be
calculated)
 % Data structures for holding solutions
t   = 0:dt:tmax;    %time
n   = length(t);    %number of time points
T   = zeros(n, 1);  %vector to hold temperature
q   = zeros(n, 1);  %vector to hold heat flux

% Set initial values
T(1)= Ti;               %initial temperature
q(1)= hc*(Tg - T(1))+e*sbc*(Tg^4 - T(1)^4);
q_coeff = dt*d/(rho*c); % A*dt/(rho*c*V); d = A/V
% loop to calculate successive values
for i=1:n-1
   T(i+1) = T(i)+q_coeff * q(i);
    q(i+1)= hc*(Tg - T(i+1))+e*sbc*(Tg^4 - T(i+1)^4);
end
 plot(t/60, T-273)
xlabel('time (min)'); ylabel('Temperature (C)')
figure
plot(t/60, q)
xlabel('time (min)'); ylabel('Heat flux (W/m^2)')
```

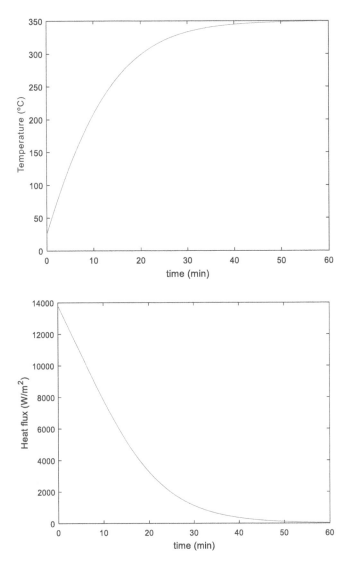

3.25 SEMI-INFINITE IDEALIZATION FOR CONCRETE AND MASONRY

The idealization of semi-infinite solids is opposite of the lumped capacity model. It is applicable to materials that are poor conductors of heat, such as concrete and masonry. Due to being poor conductors, the temperature rise away from the exposed surface of the material is gradual. It becomes possible to specify a 'depth of ingress' beyond which the rise in temperature can be neglected. Mathematically, for a 1D problem, if the exposed surface is at $x = 0$, the semi-infinite idealization implies $T = T_i$ as $x \to \infty$, as shown in Figure 3.33. The depth of ingress depends on the thermal diffusivity, α of the material. It can be shown that a temperature change at the surface will only be noticeable up to a penetration depth of $\delta < \Delta_{0.01} = 3.6\sqrt{\alpha t}$, where the number 3.6 corresponds to a temperature rise of 1% (also signified by the subscript 0.01 on Δ.

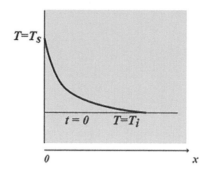

FIGURE 3.33 Semi-infinite idealization.

This relation can be established by solving the transient heat equation (3.65) with boundary conditions $T(0,t) = T_s$ (exposure temperature) and $T(\infty,t) = T_i$ (initial temperature) along with the initial condition $T(x,0) = T_i$. The solution typically proceeds through separation of variables by considering the following dimensionless variables:

$$\theta = \frac{T - T_s}{T_i - T_s}, \ X = \frac{x}{L}, \text{ and } Fo = \frac{\alpha t}{L^2} \tag{3.72}$$

where L is a characteristic length, and Fo is called Fourier number. The semi-infinite assumption is accurate for a small Fourier number; for a wall/slab of thickness $2L$, this assumption is valid for $Fo \leq 0.2$. The transient heat equation in dimensionless form can be written as

$$\frac{\partial \theta}{\partial Fo} = \frac{\partial^2 \theta}{\partial X^2} \tag{3.73}$$

with boundary conditions $\theta(0,Fo) = 0$ and $\theta(\infty,Fo) = 1$, and initial condition $\theta(X,0) = 0$. Upon solving Eq. (3.73), one obtains

$$\theta(X,Fo) = \text{erf}\left(\frac{X}{\sqrt{4Fo}} \right) \tag{3.74}$$

where $\text{erf}(z)$ denotes the Gaussian error function. In dimensional form, the solution can be written as

$$\frac{T(x,t) - T_s}{T_i - T_s} = \text{erf}\left(\frac{x}{\sqrt{4\alpha t}} \right) \tag{3.75}$$

The notion of thermal penetration depth implies that beyond a depth, $x = \delta$, the temperature rise is not substantial, as shown in Figure 3.34. δ will be a function of time and can be calculated from Eq. (3.75) for different relative rise in temperature $T(x,t)$. Penetration depths for different cases are shown in Table 3.6. The Matlab code to generate such a table is also given. It is also to be noted that the ingress depth is time-dependent (in fact, it is an increasing function of time and thermal diffusivity).

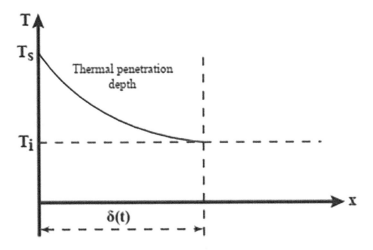

FIGURE 3.34 Thermal ingress depth.

TABLE 3.6
Thermal Ingress Depth for Various Cases

Relative Rise in Temperature $\dfrac{T-T_s}{T_i-T_s}$	$\delta(t)=2\sqrt{\alpha t}\,\mathrm{erf}^{-1}\left(\dfrac{T-T_s}{T_i-T_s}\right)$
0.99 (1% ingress of T_s)	$3.64\sqrt{\alpha t}$
0.95 (5% ingress of T_s)	$2.78\sqrt{\alpha t}$
0.90 (10% ingress of T_s)	$2.33\sqrt{\alpha t}$
0.70 (30% ingress of T_s)	$1.46\sqrt{\alpha t}$
0.50 (50% ingress of T_s)	$0.95\sqrt{\alpha t}$
0.30 (70% ingress of T_s)	$0.54\sqrt{\alpha t}$
0.10 (90% ingress of T_s)	$0.18\sqrt{\alpha t}$

ingress_depth.m
RelTRise = [0.99   0.95   0.9   0.7   0.5   0.3   0.1];
delta = 2 ª erfinv(RelTRise);

It can be seen that the expression for δ depends on the considered penetration of the surface temperature, T_s. For instance, $\dfrac{T(\delta,t)-T_s}{T_i-T_s}=0.99\Rightarrow T(\delta,t)=0.99T_i+0.01T_s$, i.e., the temperature at depth δ at time t comprises of 99% initial temperature and 1% surface temperature.

Example 3.5

The surface temperature of a thick concrete wall with an initial temperature of 0°C rises suddenly to 1,000°C.

    a. *What is the 1% thermal penetration depth δ0.01 after 15 min?*
    b. *What is the temperature T at that point after 60 min?*

Assume constant concrete properties as c = 900 Ws/(kgK), ρ = 2300 kg/m3, and k = 1.5 W/(mK).
    Given data:

$$k = \frac{1.5\,\text{W}}{\text{m}\cdot\text{K}} = 1.5\,\text{W/m}\cdot\text{K}$$

$$c = \frac{900\,\text{W}\cdot\text{s}}{\text{kg}\cdot\text{K}} = 900\,\text{W}\cdot\text{s/kg}\cdot\text{K}$$

$$\rho = \frac{2,300\,\text{kg}}{\text{m}^3} = 2,300\,\text{kg/m}^3$$

$$T_i = 20°C$$

$$T_o = 1,000°C$$

$$t_1 = 15\cdot60\,\text{s} = 900\,\text{s}$$

$$t_2 = 60\cdot60\,\text{s} = 3,600\,\text{s}$$

Calculate 1% Thermal penetration depth after 5 min

$$\alpha = \frac{k}{\rho\cdot c} = \frac{1.5\,\text{W/m}\cdot\text{K}}{2,300\,\text{kg/m}^3\cdot900\,\text{W}\cdot\text{s/kg}\cdot\text{K}} = 7.25\times10^{-07}\,\text{m}^2/\text{s}$$

$$\delta_{0.01} = 3.6\cdot\sqrt{\alpha\cdot t_1} = 3.6\cdot\sqrt{7.25\times10^{-07}\,\text{m}^2/\text{s}\cdot900\,\text{s}} = 0.0919\,\text{m}$$

Calculate temperature T after 60 min

$$\eta = \frac{\delta_{0.01}}{2\cdot\sqrt{\alpha\cdot t_2}} = \frac{0.0919\,\text{m}}{2\cdot\sqrt{7.25\times10^{-07}\,\text{m}^2/\text{s}\cdot3,600\,\text{s}}} = 0.9$$

$$\frac{T-T_o}{T_i-T_o} = \text{erf}(0.9) \Rightarrow \frac{T-1,000}{20-1,000} = 0.7969 \Rightarrow T = 219°C$$

3.26 STRUCTURAL TEMPERATURES UNDER STANDARD FIRE

Up to now, temperature calculations under simplified conditions have been discussed. In a fire scenario, typically, the fire temperature T_f may be known as a function of time, and the temperature rise within the structural members will be required to be calculated. As such, the fire temperature will be applied as a Robin boundary condition, discussed earlier, with effects of both convection and radiation (see Figure 3.31). The assumptions of lumped capacity (for metal structures) and semi-infinite idealization (for concrete/masonry structures) enable simplified expressions for the calculation of structural temperatures, especially when the structures are exposed to a standard fire. Analysis under exposure to standard fire conditions is carried out routinely to assess the fire rating of the structural member.

3.26.1 Temperature Calculation for Steel Structural Members

Temperature calculations for steel members can be carried out through the lumped capacity model introduced earlier. The governing equation is (3.69), whereas the numerical scheme is given by Eq. (3.70). The heat flux to be used is summarized in Table 3.5. The section factor A/V can be interpreted as the ratio of exposed surface area to volume (3D approach) or as exposed perimeter to area (2D approach).

Example 3.6

A steel beam carrying a floor is exposed to standard fire from three sides, as shown. Calculate its temperature as a function of time if it was at room temperature initially. The section used is ISMB 350 and assume a convective heat transfer coefficient of $h_c = 25$ W/$\left(m^2K\right)$.

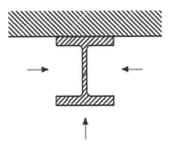

First, the required section properties are obtained and calculated.
   Given data
   Sectional properties

$h = 0.35\,m$

$b_f = 0.14\,m$

$t_f = 0.0142\,m$

$t_w = 0.0081\,m$

section factor (heated perimeter/sectional area)

$$A = 2 \cdot (h - 2 \cdot t_w) + b_f + 4 \cdot t_f + 2 \cdot (b_f - t_w) = 2 \cdot (0.35\,\text{m} - 2 \cdot 0.0081\,\text{m})$$

$$+ 0.14\,\text{m} + 4 \cdot 0.0142\,\text{m} + 2 \cdot (0.14\,\text{m} - 0.0081\,\text{m}) = 1.13\,\text{m}$$

$$V = 2 \cdot b_f \cdot t_f + t_w \cdot (h - 2 \cdot t_w) = 2 \cdot 0.14\,\text{m} \cdot 0.0142\,\text{m} + 0.0081\,\text{m} \cdot (0.35\,\text{m} - 2 \cdot 0.0081\,\text{m}) = 0.00668\,\text{m}^2$$

$$d = \frac{A}{V} = \frac{1.13\,\text{m}}{0.00668\,\text{m}^2} = 168.9\,\text{m}^{-1}$$

It is to be noted that for the calculation of A (heated perimeter), the top surface of the flange is excluded as it is not directly exposed to the fire.

```
ex_temp_steel_unprotected.m
% Given data:
d  = 168.9;    %1/m (= A/V)
Ti = 25+273;   %K (initial temperature)
hc = 25;       %W/(m^2-K)
e  = 0.9;      %emissivity
 % Assumed/found data
rho = 7850;    %kg/m^3    (density)
c   = 460;     %J/(kg-K)  (specific heat)
sbc = 5.67e-8; %W/(m^2-K^4) (Stefan-Boltzmann constant)
dt  = 10;      %s (chosen time increment)
tmax= 3600;    %s (time up to which temperatures are to be
calculated)
% Data structures for holding solutions
t  = 0:dt:tmax;    %time
n  = length(t);    %number of time points
T  = zeros(n, 1);  %vector to hold temperature
q  = zeros(n, 1);  %vector to hold heat flux
 % Gas temperature - ISO 834
Tg = 273+25+345*log10(480*(t/3600)+1);
 % Set initial values
T(1)= Ti;                  %initial temperature
q(1)= hc*(Tg(1) - T(1))+e*sbc*(Tg(1)^4 - T(1)^4);
q_coeff = dt*d/(rho*c);  % A*dt/(rho*c*V); d = A/V
 % loop to calculate successive values
for i=1:n-1
    T(i+1) = T(i) +q_coeff * q(i);
    q(i+1)= hc*(Tg(i+1) - T(i+1))+e*sbc*(Tg(i+1)^4 - T(i+1)^4);
end
 plot(t/60, Tg-273)
hold all
plot(t/60, T-273)
xlabel('time (min)'); ylabel('Temperature (C)')
legend('ISO 834', 'Steel', 'location', 'se')
```

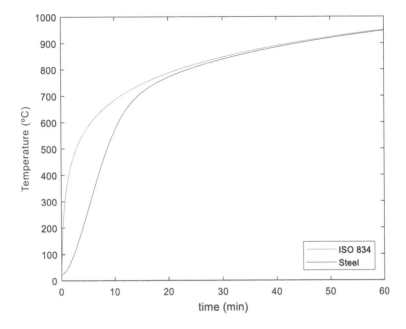

In case of protected steel members, the fire temperature T_g acts on the surface of the insulation, and the heat transfer to steel happens through the conductive mechanism. Thus, the total heat flux to be used in Eq. (3.70) becomes

$$\dot{q}'' = R_{in}\left(T_g - T\right) = \frac{d_{in}}{k_{in}}\left(T_g - T\right)$$

(3.76)

where $R_{in} = k_{in}/d_{in}$ is the conductive thermal resistance of the fire protection (k_{in} and d_{in} being the thermal conductivity and thickness of the protection material. For light insulations where the thermal storage capacity of the insulation is negligible with respect to that of steel, Eq. (3.76) can be used in Eq. (3.70) for temperature calculation as the energy balance given in Eq. (3.69) as

$$T^{(n+1)} = T^{(n)} + \Delta t \frac{A}{\rho c^{(n)} V} \frac{d_{in}}{k_{in}}\left(T_g^{(n+1)} - T^{(n)}\right)$$

(3.77)

It is to be noted that the heated perimeter A has to be computed based on the exposed perimeter of the fire protection, whereas the cross-sectional area V of the steel section is to be considered. When the heat storage capacity of fire protection material is not negligible, the energy balance has to consider the heat stored in the insulation material as well. In such a case, the temperature evolution can be calculated as

$$T^{(n+1)} = T^{(n)} + \Delta t \frac{\left(T_g^{(n+1)} - T^{(n)}\right)}{\tau\left(1 + \frac{\mu}{3}\right)} + \left(e^{\mu/\eta} - 1\right)\left(T_g^{(n+1)} - T^{(n)}\right)$$

(3.78)

with

$$\mu = \frac{A d_{in} \rho_{in} c_{in}}{V \rho c} \text{ and } \tau = \frac{\rho c V}{A} \frac{d_{in}}{k_{in}} \qquad (3.79)$$

μ is the ratio of the heat capacity of insulation to that of steel, and τ is the time constant. Notice that when $\mu \to 0$, Eqs. (3.78) becomes Eq. (3.77). η is an empirical parameter; for steel sections exposed to standard fire, $\eta = 5$ is typically used (this value is based on a large number of finite element simulations). $\eta = 10$ is used when the temperature rise due to fire is considered instantaneous; this is also the value adopted by Eurocode 3 EN 1993-1-2. Insulations having $\mu > 0.5$ are considered to have substantial heat capacity and in such case, Eq. (3.78) should be used for temperature calculations; in other scenarios, Eq. (3.77) can be used.

Example 3.7

A steel column protected by a 30 mm non-combustible board with a conductivity of 0.1 W/(m-K) is exposed to standard fire from four sides, as shown. Calculate its temperature as a function of time if it was at room temperature initially. The section used is ISHB 400 and assume a convective heat transfer coefficient of $h_c = 25$ W/(m²K).

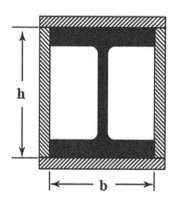

Given data
  Sectional properties

$h = 0.4m$

$b_f = 0.25$ m

$t_f = 0.0127$ m

$t_w = 0.0091$ m

  Section factor

$A = 2 \cdot (h + b_f) = 2 \cdot (0.4 \text{m} + 0.25 \text{m}) = 1.3 \text{m}$

$$V = 2 \cdot b_f \cdot t_f + t_w \cdot (h - 2 \cdot t_w) = 2 \cdot 0.25\,\text{m} \cdot 0.0127\,\text{m} + 0.0091\,\text{m} \cdot (0.4\,\text{m} - 2 \cdot 0.0091\,\text{m})$$

$$= 0.00982\,\text{m}^2$$

$$d = \frac{A}{V} = \frac{1.3\,\text{m}}{0.00982\,\text{m}^2} = 132.32\,\text{m}^{-1}$$

It is to be noted that for the calculation of A (heated perimeter), interior perimeter of the board protection is considered. Calculations are performed for both unprotected and protected versions.

ex_temp_steel_protected.m

```
% Given data:
du = 157.38;   %1/m
dp = 132.32;   %1/m (= A/V)
Ti = 25+273;   %K (initial temperature)
hc = 25;       %W/(m^2-K)
e  = 0.9;      %emissivity
din = 0.03;    %m
kin = 0.1;     %W/(m-K)
% Assumed/found data
rho = 7850;    %kg/m^3   (density)
c   = 460;     %J/(kg-K)  (specific heat)
sbc = 5.67e-8; %W/(m^2-K^4) (Stefan-Boltzmann constant)
dt  = 10;      %s (chosen time increment)
tmax= 3600;    %s (time up to which temperatures are to be
calculated)
 % Data structures for holding solutions
t   = 0:dt:tmax;    %time
n   = length(t);    %number of time points
Tu  = zeros(n, 1);  %vector to hold temperature (unprotected)
Tp  = zeros(n, 1);  %vector to hold temperature (protected)
qu  = zeros(n, 1);  %vector to hold heat flux
qp  = zeros(n, 1);  %vector to hold heat flux
% Gas temperature - ISO 834
Tg = 273+25+345*log10(480*(t/3600)+1);
 % Set initial values
Tu(1) = Ti;             %initial temperature
Tp(1) = Ti;
qu(1)= hc*(Tg(1) - Tu(1))+e*sbc*(Tg(1)^4 - Tu(1)^4);
qp(1)= hc*(Tg(1) - Tp(1))+e*sbc*(Tg(1)^4 - Tp(1)^4);
q_coeff_u = dt*du/(rho*c);
q_coeff_p = (din/kin)*dt*dp/(rho*c);
% loop to calculate successive values
for i=1:n-1
    %unprotected
    Tu(i+1) = Tu(i)+q_coeff_u * qu(i);
    qu(i+1)= hc*(Tg(i+1) - Tu(i+1))+e*sbc*(Tg(i+1)^4
 - Tu(i+1)^4);
    %protected
    Tp(i+1) = Tp(i)+q_coeff_p * qp(i);
    qp(i+1)= hc*(Tg(i+1) - Tp(i+1))+e*sbc*(Tg(i+1)^4
 - Tp(i+1)^4);
end
```

```
plot(t/60, Tg-273)
hold all
plot(t/60, Tu-273)
plot(t/60, Tp-273)
xlabel('time (min)'); ylabel('Temperature (C)')
legend('ISO 834', 'Unprotected', 'Protected', 'location', 'se')
```

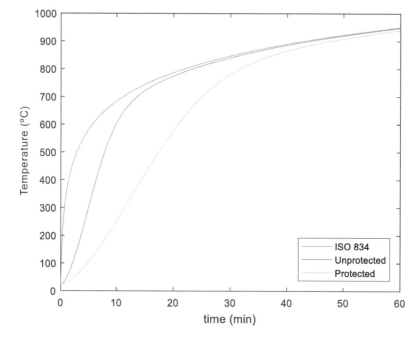

It can be observed that the section factor A/V is an important parameter that governs the temperature rise within steel members. For ease of calculations, it is beneficial to write a computer program that can automatically produce the section factor for typically used steel sections under four- and three-sided fire exposure. Table 3.7 summarizes the heated perimeter A of some frequently occurring cases, and `section_factor.m` provides a Matlab function to calculate section factors.

3.26.2   Temperature Calculation for Concrete Structural Members

The simplified approach for concrete members subjected to the ISO 834 standard fire is primarily due to Wickstrom (Wickstrom, 1985, 1986). For a 1D concrete wall (or slab) exposed to fire on one side, as shown in Figure 3.35, the approximation proceeds in a manner qualitatively similar to the idea of considering convective/radiative and conductive effects through an equivalent thermal resistance.

The fire temperature $T_f(t)$ will be known; it is assumed that a function n_s denotes the ratio of fire to surface temperature (called the surface ratio), and a function n_x denotes the ratio of surface to inside temperature (called the depth ratio), i.e.,

$$n_s = \frac{T_s}{T_f} \text{ and } n_x = \frac{T_x}{T_s} \qquad (3.80)$$

TABLE 3.7
Heated Perimeters (A) of Some Typical Exposure Case

Four-Sided Exposure   **Three-Sided Exposure**

I Shaped Section

$$A = 2D + 4B - 2t$$

$$A = 2D + 3B - 2t$$

2 Boxed Sections

$$A = 2D + 2B$$

$$A = 2D + B$$

3 Channel Sections

$$A = 2D + 4B - 2t$$

$$A = 2D + 3B - 2t$$

4 Circular Sections

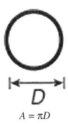

$$A = \pi D$$

5 Angle Sections

$$A = 2D + 2B$$

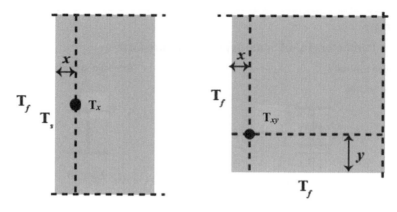

FIGURE 3.35 Calculation approach for 1D and 2D concrete elements.

section_factor.m

```
function [sf, A, V] = section_factor(section, shape,
exposure)
% Computes heated perimeter, volume and section factor.
%
% sf: section factor = A/V (1/m)
% A : heated perimeter (m)
% V : cross-sectional area (m^2)
%
% section : section dimensions; struct with fields: D, B, t
(m)
% D, B and t have meanings as defined in the figure
% t refers to web thickness
% give tf where flange thickness is needed (e.g., for I
section)
% shape : 'I', 'boxed', 'channel', 'circular', or 'angle'
% exposure: 3 or 4 for three or four sided fire exposure
switch shape
case 'I'
if exposure == 4
A = 2*section.D+4*section.B -2*section.t;
else
A = 2*section.D+3*section.B -2*section.t;
end
V = 2*section.B*section.tf+(section.D -2*section.
tf)*section.t;
case 'boxed'
if exposure == 4
A = 2*section.D+2*section.B;
else
A = 2*section.D+section.B;
end
V = section.B*section.D -...
(section.B -2*section.t)*(section.D -2*section.t);
```

```
case 'channel'
if exposure == 4
A = 2*section.D+4*section.B -2*section.t;
else
A = 2*section.D+3*section.B -2*section.t;
end
V = 2*section.B*section.tf+(section.D -2*
section.tf)*section.t;
case 'circular'
A = pi*section.D;
V = (pi*section.D^2- pi*(section.D - section.t)^2)/4;
case 'angle'
% note that this is implemented for 4 sided exposure only
A = 2*section.D+2*section.B;
V = section.t*(section.B+section.D - section.t);
otherwise
error('Unknown shape');
end
sf = A/V;
```

such that the temperature at a depth x within the concrete member can be written as

$$T_x(x,t) = n_s(t)n_x(x,t)T_f(t) \tag{3.81}$$

Considering a large number of scenarios (generated through computer simulations), the forms of n_s and n_x have been determined to be

$$n_s = 1 - 0.060t^{-0.90} \tag{3.82}$$

and

$$n_x = 0.172\ln\left(\frac{t}{x^2}\right) - 0.74 \tag{3.83}$$

where time t is to be provided in h, and distance x is to be provided in meters. When using the ISO standard fire, $T_f(t) = 345\log(480t + 1)$ can be used. The aforementioned expressions for n_s and n_x are valid for normal weight concrete having water content of 1.5% and have been adopted by Eurocode 2.

Example 3.8

A 230 mm thick concrete wall is subjected to a one-sided ISO 834 fire exposure. Determine the temperature within the concrete wall at different depths for a 4 h exposure. Assume the concrete is of normal weight.

ex_temp_conc_1D.m

```
% Given data
ts = 0.1:0.1:4;              %time up to 4 hours
xs = [1 5 10 15 20]/1e3;     %depths in m
Tx = zeros(length(xs), length(ts));
Tf = 20+345*log10(480*ts+1);
 for x = 1:length(xs)
    for t = 1:length(ts)
        ns = 1 - 0.06*ts(t)^(-0.9);
        nx = 0.172*log(ts(t)/xs(x)^2) - 0.74;
        Tx(x,t) = ns*nx*Tf(t);
    end
end
figure; hold all
plot(ts, Tf);
for x = 1:length(xs)
    plot(ts, Tx(x,:));
end
```

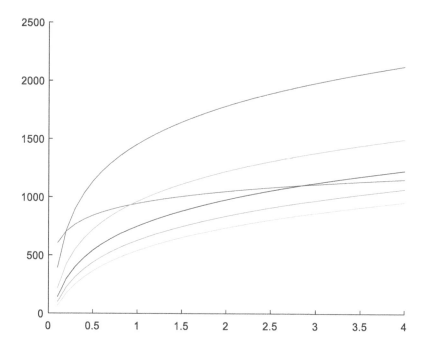

For members exposed on two sides (e.g., a beam or a column), the same idea is extended as below (see 36).

$$T_{xy}(x,y,t) = \left[n_s \left(n_x + n_y - 2n_x n_y \right) + n_x n_y \right] T_f \qquad (3.84)$$

with n_s and n_x being the same as defined earlier in Eqs. (3.82) and (3.83), respectively, and n_y being defined analogously as

$$n_y = 0.172 \ln\left(\frac{t}{y^2}\right) - 0.74 \qquad (3.85)$$

Example 3.9

Calculate the temperature in a rectangular concrete beam (depth 450 mm, width 230 mm) after 1h fire exposure at a point 30 mm and 25 mm from the exposed surfaces. Also plot its temperature contours after 1 h and variation of temperature at the said point with time.

Given data (time in h, distances in m):

$t = 1\,h$
$x = 0.03\,m$
$y = 0.025\,m$

Calculations

$$n_s = 1 - 0.06 \cdot t^{-0.9} = 1 - 0.06 \cdot 1^{-0.9} = 0.94$$

$$n_x = 0.172 \cdot ln\left(\frac{t}{x^2}\right) - 0.74 = 0.172 \cdot ln\left(\frac{1}{0.03^2}\right) - 0.74 = 0.466$$

$$n_y = 0.172 \cdot ln\left(\frac{t}{y^2}\right) - 0.74 = 0.172 \cdot ln\left(\frac{1}{0.025^2}\right) - 0.74 = 0.529$$

$$T_f = 345 \cdot log(480 \cdot t + 1) = 345 \cdot log(480 \cdot 1 + 1) = 925.34\,°C$$

$$T_{xy} = \left(n_s \cdot (n_x + n_y - 2 \cdot n_x \cdot n_y) + n_x \cdot n_y\right) \cdot T_f$$

$$= \left(0.94 \cdot (0.466 + 0.529 - 2 \cdot 0.466 \cdot 0.529) + 0.466 \cdot 0.529\right) \cdot 925.34 = 664.83\,°C$$

Contour plot of the lower quarter portion of the beam is shown below. The contour of 500°C (or 500°C isotherm is plotted with emphasis). Note that only half width of the beam is shown; the temperature distribution will be symmetric about the central vertical axis.

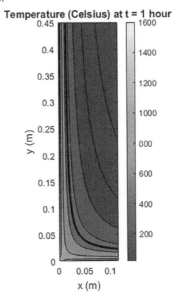

Variation of temperature of (0.03, 0.025) point with time is shown below.

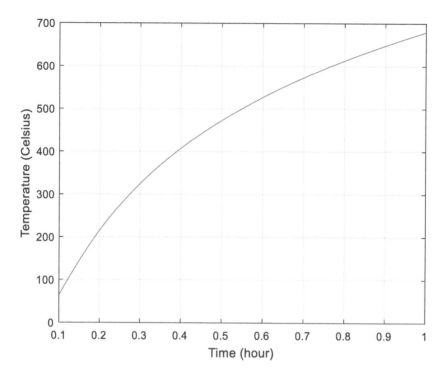

Matlab code used for calculations to generate these graphs is given below.

```
ex_temp_conc_2D.m
% Given data
ts = 0.1:0.01:1;                   %time up to 4 hours
xs = [0:1:230]*1e-3;               %width, m
ys = [0:1:450]*1e-3/2;             %depth, m
[X, Y] = ndgrid(xs, ys);
Txy = zeros(length(xs), length(ys), length(ts));
Tf = 20+345*log10(480*ts+1);
 ns = 1 - 0.06*ts.^(-0.9);
 for t=1:length(ts)
   nx = 0.172*log(ts(t)./(X.^2)) - 0.74;
   ny = 0.172*log(ts(t)./(Y.^2)) - 0.74;
   Txy(:, :, t) = (ns(t)*(nx+ny - 2*nx.*ny)+nx.*ny)*Tf(t);
end
 figure, contourf(X, Y, Txy(:, :, end)); colorbar
xlabel('x (m)'); ylabel('y (m)');
title('Temperature (Celsius) at t = 1 hour')
 Tpoint_1hour = griddata(X, Y, Txy(:, :, end), 0.03, 0.025);
Tpoint = zeros(length(ns), 1);
for t=1:length(ns)
   Tpoint(t) = griddata(X, Y, Txy(:, :, t), 0.03, 0.025);
end
figure, plot(ts, Tpoint); grid on; box on;
xlabel('Time (hour)'); ylabel('Temperature (Celsius)')
```

3.27   DESIGN THEORY – LIMIT STATES OF DESIGN

A limit state function in design is typically defined as

$$L(D, R) = R - D \geq 0 \tag{3.86}$$

where D and R denote the load demand and the resistance of the system, respectively. The load demand, D, may arise from a variety of load cases (or their combinations), such as dead loads, live or imposed loads, and wind loads. The resistance, R, is the carrying capacity of the structure defined with respect refers to relevant action. For instance, it can be the axial force capacity for a member under tension or compression, bending moment capacity for a member under bending, or a combination. In the present context, it will be beneficial to introduce different notations for the demand and resistance in normal (cold) conditions and hot (fire) conditions as D_{cold} and R_{cold}, and D_{hot} and R_{hot}, respectively.

The load demand in a fire scenario typically (and ideally) reduces with time due to the egress of occupants and burning of combustibles in the space, i.e., typically $D_{hot} \leq D_{cold}$. However, lower load demand need not necessarily imply lower stresses within the structural members as fire can induce significant additional stresses due to thermal expansion (especially for indeterminate structures). Most codes (ASCE, New Zealand, and Australia) consider dead and live loads for considering the load demand during a fire scenario. Eurocode provides additional load combinations considering the actions of wind and snow. A brief summary of such load combinations is given in Eq. (3.87). Here, DL, PLL, and VLL denote dead load, permanent live load, and variable live load, respectively.

$$\begin{array}{ll} .0 \text{ DL} + 0.9 \text{ PLL} + 0.5 \text{ VLL} & \text{Eurocode} \\ 1.2 \text{ DL} + 0.5 \text{ PLL} + 0.5 \text{ VLL} & \text{ASCE} \\ 1.0 \text{ DL} + 0.6 \text{ PLL} + 0.4 \text{ VLL} & \text{New Zealand, Australia} \end{array} \tag{3.87}$$

The load resistance R of structural members usually decreases with increase in temperate due to degradation in the material strength, as discussed earlier. The overall safety of a structural system during fire requires

$$R_{fire} \geq D_{fire} \tag{3.88}$$

where R_{fire} and D_{fire} are equal to R_{cold} and D_{cold} at the beginning of the fire event and decrease with time as the fire continues. The time at which the equality holds in Eq. (3.88) is the strength fire rating of the structural system (typically expressed in hours or minutes).

When compared to normal temperature design, design under a fire scenario typically entails: (i) lower amount of applied loads; (ii) possibility of additional internal forces or displacements due to thermal expansion; (iii) reduction in strength of materials; (iv) reduction in cross-sectional areas, e.g., due to spalling in case of concrete; and (v) possibility of different failure mechanisms – for instance, the location of plastic hinges may change. Additionally, design under fire usually involves smaller values of the safety factors due to the less frequency of the event.

A summary list of relevant codes and standards pertaining to design under fire conditions is presented next.

1. American codes
    a. NFPA 5000 – building construction and safety code.
    b. ACI 216 – design of concrete structures
    c. AISC Design guide 19 – design of steel structures
    d. International Building code (IBC) 2015 – for different applications
    e. National Design Specification (NDS) for Wood Construction – timber structures
2. Australian codes
    a. Building Code of Australia (BCA)
    b. AS 1170.x – structural design actions
    c. AS 4100 – design of steel structures
    d. AS 3600 – design of concrete structures
    e. AS 1720 – design of timber structures
3. British codes
    a. BS 8110 – design of concrete structures
    b. BS 5268 – design of timber structures
    c. BS 5268 – design of steel structures
4. Canadian codes
    a. National Building Code of Canada (NBCC)
    b. CAN/CSA-A23.3 – design of concrete structures
    c. S16-14 – design of steel structures
    d. O86-14 – engineering design in wood
5. Eurocodes
    a. Eurocode 1 – design basis and actions
    b. Eurocode 2 – design of concrete structures
    c. Eurocode 3 – design of steel structures
    d. Eurocode 4 – design of composite steel and concrete structures
    e. Eurocode 5 – design of timber structures
    f. Eurocode 6 – design of masonry structures
    g. Eurocode 9 – design of aluminum structures
6. Indian codes
    a. National Building Code (NBC) Part 4 – design for fire and life safety
7. New Zealand codes
    a. Standards New Zealand (SNZ)
    b. NZS 4203 – general design of structures and load specifications
    c. NZS 3101 – design of concrete structures
    d. NZS 3404 – design of steel structures
    e. NZS 3603 – design of timber structures

3.28   TENSION MEMBERS

For tension members, the resistance R refers to the axial load carrying capacity, N, of the member. As steel is usually used for tension members, the design examples in this section pertain to design of steel members under tension. As has been discussed earlier, the Young's modulus, proportional limit, and yield strength of steel decrease with increase in temperature with k_{ET}, k_{pT}, and k_{yT} being the corresponding reduction factors. The design resistance fire conditions can be calculated as

$$N_{fire} = k_{yT} f_y A_{fire} \tag{3.89}$$

where f_y is the normal temperature yield strength of the material and A_{fire} is the area of cross section. Usually, $A_{fire} = A_{cold}$ for steel sections, but A_{fire} allows the consideration of scenarios where the cross-sectional area may reduce during a fire event. If $N_{cold} = f_y A_{cold}$, one may view the capacity under fire conditions as $N_{fire} = k_{yT} N_{cold}$. It is to be noted that there will be situations (e.g., when bolted connections are used) where a material stress other than the yield strength will be required. The methodology, however, will not change and the corresponding strength reduction factor can be used in a similar manner.

Example 3.10

An angle section $100 \times 75 \times 6$ is used as a tension member in a roof truss. Calculate its design strength in gross-section yielding under cold conditions. If the member is exposed to a standard fire on all four sides, calculate its temperature and strength as a function of time.

```
ex_steel_tension1.m
% section properties
section.B = 75e-3;     %m
section.D = 100e-3;    %m
section.t = 6e-3;      %m
[sf, Am, V] = section_factor(section, 'angle', 4);
 % material data:
hc = 25;        %W/(m^2-K)  (convection coefficient)
e  = 0.9;       %emissivity
fy = 250e6;     %Pa   (yield strength)
% Assumed/found data
rho = 7850;     %kg/m^3    (density)
c   = 460;      %J/(kg-K)  (specific heat)
sbc = 5.67e-8;  %W/(m^2-K^4) (Stefan-Boltzmann constant)
dt  = 10;       %s (chosen time increment)
tmax= 3600;     %s (time up to which temperatures are to be
calculated)
 % Data structures for holding solutions
t   = 0:dt:tmax;      %time
n   = length(t);      %number of time points
T   = zeros(n, 1);    %vector to hold temperature
```

```
q   = zeros(n, 1);  %vector to hold heat flux
Ti = 25+273;  %K (initial temperature)
% Gas temperature - ISO 834
Tg = 273+25+345*log10(480*(t/3600)+1);
 % Set initial values
T(1)= Ti;              %initial temperature
q(1)= hc*(Tg(1) - T(1))+e*sbc*(Tg(1)^4 - T(1)^4);
q_coeff = dt*sf/(rho*c); % A*dt/(rho*c*V); d = A/V
 % loop to calculate successive values of temperature
for i=1:n-1
   T(i+1) = T(i)+q_coeff * q(i);
   q(i+1)= hc*(Tg(i+1) - T(i+1))+e*sbc*(Tg(i+1)^4 - T(i+1)^4);
end
 % calculate strength reduction factor
addpath('./mat_props/');
ky = steel_yield_strength_reduction(T);
 % calculate reduced strength with respect to time
% use appropriate partial safety factor for material strength
% V is the area of cross-section
Pcold = fy * V;
Phot  = ky * Pcold;
 % plot
plot(t/60, Tg-273)
hold all; grid on; box on
plot(t/60, T-273)
xlabel('Time (min)'); ylabel('Temperature (Celsius)')
legend('Gas', 'Member')
 figure
plot(t/60, Phot/1000); grid on; box on
xlabel('Time (min)'); ylabel('Strength (kN)')
```

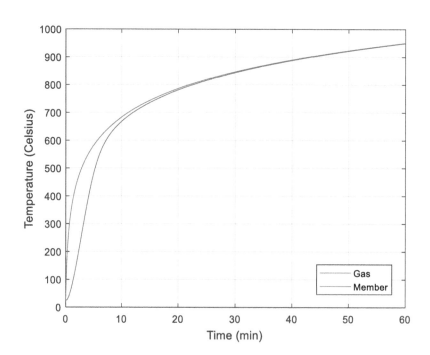

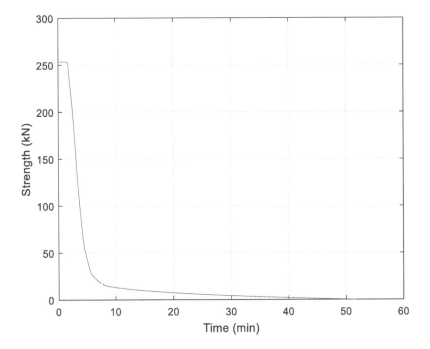

Example 3.11

An ISLB 250 section is used as a tension member. Though its end connections are protected against fire, the middle portion is unprotected. Calculate its load resistance a function of time if it is exposed to an ISO 834 fire.

Since the ends are protected, the middle portion will become critical during fire. Further, calculation of temperature evolution is necessary to ascertain the load resistance.

```
ex_steel_tension2.m
addpath('./mat_props/');
% section properties
section.B = 125e-3;      %m
section.D = 250e-3;      %m
section.t = 6.1e-3;      %m
section.tf= 8.2e-3;      %m
[sf, Am, V] = section_factor(section, 'I', 4);
% material data:
hc = 25;       %W/(m^2-K)  (convection coefficient)
e  = 0.9;      %emissivity
fy = 250e6;    %Pa   (yield strength)
 % Assumed/found data
rho = 7850;    %kg/m^3    (density)
c   = 460;     %J/(kg-K)  (specific heat)
sbc = 5.67e-8; %W/(m^2-K^4) (Stefan-Boltzmann constant)
dt  = 10;      %s (chosen time increment)
tmax= 3600;    %s (time up to which temperatures are to be
calculated)
 % Data structures for holding solutions
```

```
t   = 0:dt:tmax;     %time
n   = length(t);     %number of time points
T   = zeros(n, 1);   %vector to hold temperature
q   = zeros(n, 1);   %vector to hold heat flux
 Ti = 25+273;   %K (initial temperature)
% Gas temperature - ISO 834
Tg = 273+25+345*log10(480*(t/3600)+1);
% Set initial values
T(1)= Ti;              %initial temperature
q(1)= hc*(Tg(1) - T(1))+e*sbc*(Tg(1)^4 - T(1)^4);
% loop to calculate successive values of temperature
for i=1:n-1
    c = steel_specific_heat(T(i)); %temperature-dependent specific
heat
    q_coeff = dt*sf/(rho*c); % A*dt/(rho*c*V); d = A/V
    T(i+1) = T(i)+q_coeff * q(i);
    q(i+1)= hc*(Tg(i+1) - T(i+1))+e*sbc*(Tg(i+1)^4 - T(i+1)^4);
end
 % calculate strength reduction factor
ky = steel_yield_strength_reduction(T);
 % calculate reduced strength with respect to time
% use appropriate partial safety factor for material strength
% V is the area of cross-section
Pcold = fy * V;
Phot  = ky * Pcold;
 % plot
plot(t/60, Tg-273)
hold all; grid on; box on
plot(t/60, T-273)
xlabel('Time (min)'); ylabel('Temperature (Celsius)')
legend('Gas', 'Member')
 figure
plot(t/60, Phot/1000); grid on; box on
xlabel('Time (min)'); ylabel('Strength (kN)')
```

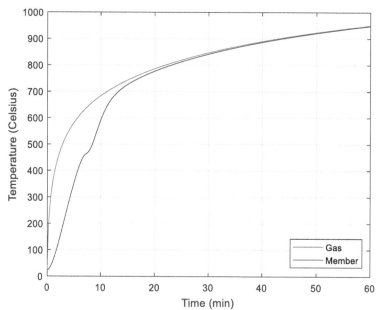

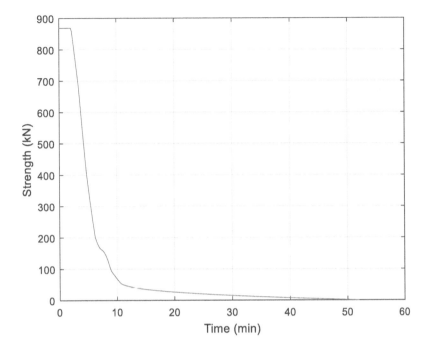

Compare the results with Example 3.11 and notice that although the absolute strength is greater in the current example, the degradation of strength is similar and the tension members in both the examples are not expected to last more than 10 min in case of a fire (of course, it depends on the load demand as well). Also notice that a key difference in temperature calculations in this example vs. the previous one is that this example uses the Eurocode temperature-dependent model for specific heat of steel.

The readers are encouraged to repeat these examples with different configurations of fire protection applied to the members and compare the strength reduction with unprotected member calculations. A similar Matlab code can be developed for this purpose; only that the temperature calculations will be required to be done considering the member insulations. It is also to be noted that these calculations consider the lumped capacitance model for thermal analysis, which is reasonable in most of the scenario (and usually conservative in others). In case of significantly non-uniform structural temperatures, one may consider the section to be comprising of several smaller portions of uniform temperature (typically based on temperature contours) and carry out the calculations accordingly. In such cases, a more detailed thermal analysis is also required, which proceeds through the use of finite element models.

Another aspect to consider in temperature analysis of steel members in which certain portions of the member can shield other portions from radiative flux from the fire (e.g., flange of an I-section shielding part of its web). Eurocode suggests the use of a shielding factor, k_{sh}, in the calculation of radiative heat flux (in lieu of the configuration factor) as

$$\dot{q}_r'' = k_{sh} \epsilon \sigma \left(T_g^4 - T_s^4 \right) \tag{3.90}$$

The configuration factor is calculated as

$$k_{sh} = 0.9\frac{\left[\dfrac{A}{V}\right]_b}{\dfrac{A}{V}}$$

(3.91)

where $\dfrac{A}{V}$ is the section factor of the actual section, while $\left[\dfrac{A}{V}\right]_b$ is the section factor calculated by considering the box perimeter instead of the actual section perimeter A.

3.29 DESIGN FOR FLEXURE – BEAMS

For beams, the resistance R primarily refers to the bending moment capacity, M, of the member. In many cases, one needs to also consider the shear resistance, V, as well. Design of steel and reinforced beams require different considerations and are covered in the subsequent sections.

3.29.1 STEEL BEAMS

Design of steel beams follows a similar process as that of tension members. For beams where lateral torsional buckling is not critical, the moment-carrying capacity can be simply calculated as

$$M_{fire} = k_{yT}M_{cold}$$

(3.92)

where M_{cold} is the normal temperature moment-carrying capacity of the beam. For laterally unrestrained beams, the methodology of using reduction factors for strength and Young's modulus can be propagated in a similar manner to compute the elastic lateral torsional buckling moment as

$$M_{cr,fire} = \left[\frac{\pi^2(k_E E)I_y}{L_{LT}^2}\right]^{0.5}\left[(k_G G)I_t\,\frac{\pi^2(k_E E)I_w}{L_{LT}^2}\right]^{0.5}$$

(3.93)

where k_E and k_G denote the reduction factors for Young's modulus and shear modulus, respectively, L_{LT} is the effective length for lateral torsional buckling, I_t is the torsional constant, I_w is the warping constant, and I_y is the moment of inertia about the weaker axis. $M_{cr,fire}$ can be used to compute the critical buckling stress at elevated temperature,

$$f_{cr,lt,fire} = M_{cr,fire}\,/\,Z_p$$

(3.94)

with Z_p being the plastic section modulus of the section; note that most codes include a reduction factor to be multiplied to consider the cases where the section is slender. The non-dimensional slenderness ratio, λ_{LT}, can then be calculated for heated conditions as

$$\lambda_{LT,fire} = \left[\frac{k_y f_y}{f_{cr,lt,fire}}\right]^{0.5} \tag{3.95}$$

It is to be noted that the reduction factors used in Eqs. (3.93) and (3.95) should correspond to the temperature of the flange of the section. In case uniform sectional temperature is assumed, the temperature will be same for the entire section, and in that case, the reduction factors will also be the same for the entire section. Next, the bending stress reduction factor to account for lateral torsional buckling at elevated temperature can be calculated as

$$\chi_{LT,fire} = \frac{1}{\phi_{LT,fire} + \left[\phi_{LT,fire}^2 - \lambda_{LT}^2\right]^{0.5}}$$

$$\phi_{LT,fire} = 0.5\left[1 + \alpha_{LT}\left(\lambda_{LT,fire} - 0.2\right) + \lambda_{LT}^2\right] \tag{3.96}$$

where α_{LT} is an imperfection factor accounting for the fabrication uncertainties. Although the aforementioned relations are from IS 800:2007, they arise from similar theoretical considerations as other building codes. Some empirical factors such as α_{LT} may be different in different codes, but the overall methodology will remain the same. The moment-carrying capacity at high temperatures can then be calculated as

$$M_{fire} = Z_p \chi_{LT,fire}\left(k_y f_y\right) \tag{3.97}$$

Example 3.12

What will be the fire rating of an unprotected steel beam built using ISMB 350 having a moment demand of 150 kNm exposed to three-sided ISO 834 fire?

```
ex_steel_beam1.m
addpath('./mat_props/');
% section properties
section.B  = 140e-3;     %m
section.D  = 350e-3;     %m
section.t  = 8.1e-3;     %m
section.tf = 14.2e-3;    %m
section.Iz = 13600e-8;   %m^4
section.Ze = 779e-6;     %m^3
section.Zp = section.Ze*1.12;
```

```
Mdemand = 150;   %kN-m
 [sf, Am, V] = section_factor(section, 'I', 3);
[sf_b, Am_b, V_b] = section_factor(section, 'boxed', 3);
ksh = 0.9*Am_b/Am;        %shielding factor
 % material data:
hc = 25;         %W/(m^2-K)   (convection coefficient)
e  = 0.7;        %emissivity
fy = 250e6;      %Pa    (yield strength)
 % Assumed/found data
rho = 7850;      %kg/m^3     (density)
c   = 460;       %J/(kg-K)   (specific heat)
sbc = 5.67e-8;   %W/(m^2-K^4)  (Stefan-Boltzmann constant)
dt  = 10;        %s (chosen time increment)
tmax= 1800;      %s (time up to which temperatures are to be
calculated)
 % Data structures for holding solutions
t   = 0:dt:tmax;     %time
n   = length(t);     %number of time points
T   = zeros(n, 1);   %vector to hold temperature
q   = zeros(n, 1);   %vector to hold heat flux
 Ti = 25+273;   %K (initial temperature)
% Gas temperature - ISO 834
Tg = 273+25+345*log10(480*(t/3600)+1);
 % Set initial values
T(1)= Ti;              %initial temperature
q(1)= hc*(Tg(1) - T(1))+ksh*e*sbc*(Tg(1)^4 - T(1)^4);
 % loop to calculate successive values of temperature
for i=1:n-1
   c = steel_specific_heat(T(i)); %temperature-dependent specific
heat
   q_coeff = dt*sf/(rho*c);        % A*dt/(rho*c*V); d = A/V
   T(i+1) = T(i)+q_coeff * q(i);
   q(i+1)= hc*(Tg(i+1) - T(i+1))+ksh*e*sbc*(Tg(i+1)^4
- T(i+1)^4);
end
 % calculate strength reduction factor
ky = steel_yield_strength_reduction(T);
 % calculate reduced strength with respect to time
% use appropriate partial safety factor for material strength
% V is the area of cross-section
Mcold = section.Zp * fy / 1.10;   %1.10 is material safety factor
Mhot  = ky * Mcold;

% plot
plot(t/60, Tg-273)
hold all; grid on; box on
plot(t/60, T-273)
line(xlim, [550 550])
xlabel('Time (min)'); ylabel('Temperature (Celsius)')
legend('Gas', 'Member', 'Critical Temperature', 'location', 'nw')
figure
plot(t/60, Mhot/1000); grid on; box on
line(xlim, [Mdemand Mdemand])
xlabel('Time (min)'); ylabel('Moment (kNm)')
legend('Moment capacity', 'Moment demand')
```

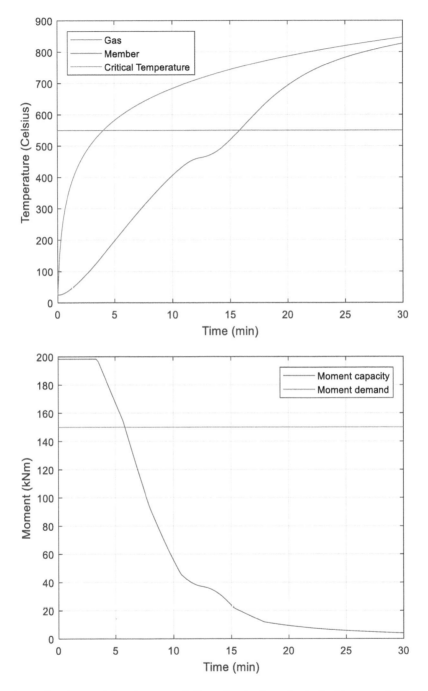

It can be observed that the strength fire rating is about 6 min. In this example, the shielding factor for radiation is also employed.

Example 3.13

Calculate the fire rating of the beam of Example 12 if it was protected with a 30mm fibreboard protection.

```
ex_steel_beam2.m
addpath('./mat_props/');
% section properties
section.B  = 140e-3;      %m
section.D  = 350e-3;      %m
section.t  = 8.1e-3;      %m
section.tf = 14.2e-3;     %m
section.Iz = 13600e-8;    %m^4
section.Ze = 779e-6;      %m^3
section.Zp = section.Ze*1.12;

Mdemand = 150;   %kN-m
 [~, Am, V] = section_factor(section, 'I', 3);
 [sf_b, Am_b, V_b] = section_factor(section, 'boxed', 3);
sf = Am_b / V;           %due to fire protection
ksh = 1;                 %shielding factor
 % insulation properties
din = 0.03;      %m
kin = 0.25;      %W/(m-K)
% material data:
hc = 25;         %W/(m^2-K)  (convection coefficient)
e  = 0.7;        %emissivity
fy = 250e6;      %Pa   (yield strength)
% Assumed/found data
rho = 7850;      %kg/m^3     (density)
c   = 460;       %J/(kg-K)   (specific heat)
sbc = 5.67e-8;   %W/(m^2-K^4) (Stefan-Boltzmann constant)
dt  = 10;        %s (chosen time increment)
tmax= 3600;      %s (time up to which temperatures are to be
calculated)
 % Data structures for holding solutions
t   = 0:dt:tmax;    %time
n   = length(t);    %number of time points
T   = zeros(n, 1);  %vector to hold temperature
q   = zeros(n, 1);  %vector to hold heat flux
 Ti = 25+273;   %K (initial temperature)
% Gas temperature - ISO 834
Tg = 273+25+345*log10(480*(t/3600)+1);
 % Set initial values
T(1)= Ti;               %initial temperature
q(1)= hc*(Tg(1) - T(1))+ksh*e*sbc*(Tg(1)^4 - T(1)^4);
 % loop to calculate successive values of temperature
for i=1:n-1
   c = steel_specific_heat(T(i)); %temperature-dependent specific
heat
   q_coeff = (din/kin)*dt*sf/(rho*c);
```

```
    T(i+1) = T(i)+q_coeff * q(i);
    q(i+1)= hc*(Tg(i+1) - T(i+1))+ksh*e*sbc*(Tg(i+1)^4
- T(i+1)^4);
end
 % calculate strength reduction factor
ky = steel_yield_strength_reduction(T);
 % calculate reduced strength with respect to time
 % use appropriate partial safety factor for material strength
 % V is the area of cross-section
Mcold = section.Zp * fy / 1.10;    %1.10 is material safety factor
Mhot  = ky * Mcold;
 % plot
plot(t/60, Tg-273)
hold all; grid on; box on
plot(t/60, T-273)
line(xlim, [550 550])
xlabel('Time (min)'); ylabel('Temperature (Celsius)')
legend('Gas', 'Member', 'Critical Temperature', 'location', 'nw')
 figure
plot(t/60, Mhot/1000); grid on; box on
xlim([0 30])
linc(xlim, [Mdemand Mdemand])
xlabel('Time (min)'); ylabel('Moment (kNm)')
legend('Moment capacity', 'Moment demand')
```

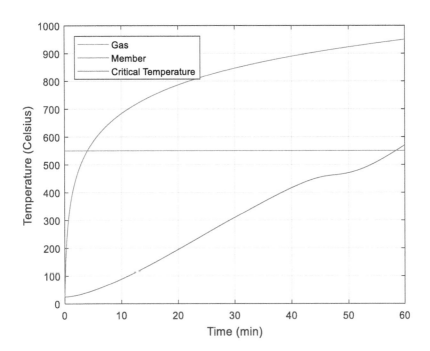

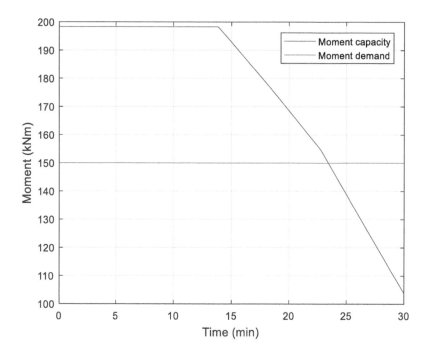

The fire rating is about 24 min. The effectiveness and importance of providing fire protection on steel members can also be seen through this example. The fire rating of the protected beam is almost four times that of the unprotected beam.

It is advantageous to have an automated mechanism to obtain the relevant material properties of insulation materials without having to look up tables in the literature. For this purpose, an indicative Matlab function is being provided as insulation_properties.m. The readers are encouraged to expand this function and use them in the calculations.

insulation_properties.m

```
function props = insulation_properties(insulation_type)
% returns relevant properties of insulation materials as a struct
(props)
% the code is self-explanatory
% everything is in the basic SI units
  switch insulation_type
    case 1
       %Gypsum plaster
       props.rho = 800;    %density, kg/m^3
       props.c   = 1700;   %specific heat, J/kg.K
       props.k   = 0.2;    %thermal conductivity, W/m.K
       props.u   = 20;     %nominal moisture content, percent
    case 2
       %Light weight concrete
       props.rho = 600;
       props.c   = 1200;
       props.k   = 0.8;
       props.u   = 2.5;
```

```
case 3
    %Normal weight concrete
    props.rho = 2200;
    props.c   = 1200;
    props.k   = 1.7;
    props.u   = 1.5;
case 4
    %Aerated concrete
    props.rho = 600;
    props.c   = 1200;
    props.k   = 0.3;
    props.u   = 2.5;
case 5
    %Mineral fiber sheet
    props.rho = 500;
    props.c   = 1500;
    props.k   = 0.25;
    props.u   = 2.0;
case 6
    %Sprayed mineral fiber
    props.rho = 300;
    props.c   = 1050;
    props.k   = 0.1;
    props.u   = 1.0;
case 7
    %Cement board
    props.rho = 1600;
    props.c   = 840;
    props.k   = 0.36;
    props.u   = 10;
case 8
    %Brick masonry
    props.rho = 1700;
    props.c   = 800;
    props.k   = 0.73;
    props.u   = 1.0;
otherwise
    error('Properties not available in database. Please
update.')
end
```

3.29.2 REINFORCED CONCRETE BEAMS

Thermal analysis of concrete members was discussed earlier through Wickstrom's method. While the earlier equations (3.81) for 1D calculations and (3.84) for 2D calculations are to be used for normal density concrete, they have been generalized to consider other kinds of concrete (e.g., lightweight) by including the thermal diffusivity in the expressions of n_x and n_y as

$$n_x = 0.172\ln\left(\frac{a}{a_c}\frac{t}{x^2}\right) - 0.74$$

$$n_y = 0.172\ln\left(\frac{a}{a_c}\frac{t}{y^2}\right) - 0.74 \tag{3.98}$$

where a is the thermal diffusivity of the material under consideration and $a_c = 0.417 \times 10^{-6}$ m^2 / s is the thermal diffusivity of normal density concrete.

The overall design of reinforced concrete beams can be carried out in a simplified manner through the 500°C isotherm method, prescribed by a few building codes. The assumptions and key considerations of the 500°C isotherm method are as follows:

- Concrete within the 500°C isotherm is considered to retain its full strength while that outside of this contour is considered to have zero strength. This simplifies the strength considerations as temperature-dependent strength reduction calculations for concrete are not required. There are methods that consider this as well, which fall under the category of zone methods (one such method is Hert'z method). This idea is shown in Figure 3.36.
- The geometrical section of the beam is reduced in accordance with the 500°C isotherm; i.e., the cold width b of the beam is reduced to b_{fire}.
- Strength calculations of rebars are carried out as per actual strength reduction depending on their actual temperature and strain. The location of the rebars is considered to remain the same

A schematic of the sectional analysis of a reinforced concrete beam is shown in Figure 3.37. The depth of the concrete stress block on compression side is taken as λx, where x is the depth of the neutral axis and λ is given as

$$\lambda = 0.8 - \frac{f_{ck} - 50}{200} \leq 0.8 \qquad (3.99)$$

with f_{ck} being the characteristic strength of concrete. Design strength of concretes taken as $f_{cd} = \eta f_{ck}$ with $\eta = 1 - \frac{f_{ck} - 50}{200} \leq 1$. The total moment-carrying capacity of the beam comprises two portions as $M_u = M_{u1} + M_{u2}$, where M_{u1} is the capacity of the balanced concrete steel section, while M_{u2} is the capacity of the compression and balancing tension reinforcement.

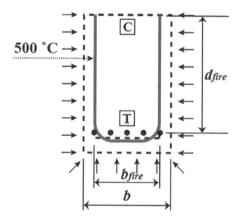

FIGURE 3.36   Consideration of b_{fire} and d_{fire} in a three-sided exposure.

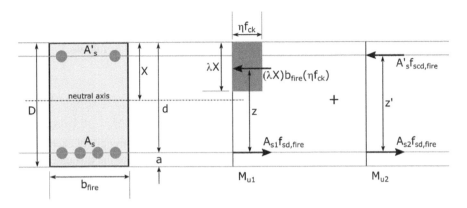

FIGURE 3.37   Sectional analysis showing stress distribution at ultimate limit state.

These moment capacities are calculated as

$$M_{u1} = A_{s1} f_{sd,fire} z$$

$$M_{u2} = A_{s2} f_{sd,fire} z' = A'_s f_{scd,fire} z' \qquad (3.100)$$

Here, A'_s is the area of the compression steel, $f_{sd,fire}$ and $f_{scd,fire}$ are the design tensile and compressive strengths of tension and compression reinforcement steel, respectively, at the average temperature of a particular reinforcement layer, and z and z' are the lever arms shown in Figure 3.37. The calculation of $f_{sd,fire}$ is discussed next; $f_{scd,fire}$ can be calculated with a similar approach. Let a reinforcement layer comprise of n rebars ($n = 4$ in Figure 3.37). The average temperature of the reinforcement layer can be calculated as

$$T_{avg} = \frac{\sum_{i=1}^{n} T_i}{n} \qquad (3.101)$$

where T_i is the temperature of each rebar found from the thermal analysis. Due to small volume of steel with respect to concrete, thermal analysis can be performed by assuming that the entire cross section comprises of concrete, and the temperatures at the locations of rebars can be considered as temperatures of the rebars without much error. Once T_{avg} is known, the average strength reduction factor of steel can be calculated as

$$k_{y,avg} = \frac{\sum_{i=1}^{n} k_{s,i}}{n} \qquad (3.102)$$

where $k_{s,i}$ is the strength reduction factor of each rebar (calculated from T_i mentioned earlier). The average design tensile strength can then be calculated as

$$f_{sd,fire} = k_{y,avg} f_{sd} \qquad (3.103)$$

with f_{sd} being the design tensile strength of reinforcing steel at 20°C. A more accurate calculation of $f_{sd,fire}$ may be needed, in case the rebars have different areas and/or are not distributed in uniform layers. This can be carried out as follows:

$$f_{sd,fire} = \frac{\sum_{i=1}^{n}\left[k_{s,i}f_{sd,i}A_i\right]}{\sum_{i=1}^{n}A_i} \tag{3.104}$$

where A_i is the area of each rebar, and $f_{sd,i}$ is the design tensile strength of each rebar.

In order to calculate the lever arms, it is required to calculate the distance of the centroid of the reinforcement layer from the edge of the reduced cross section of the beam. The generic expression to calculate this centroid distance (in line with Eq. (3.104)) is given by

$$a = \frac{\sum_{i=1}^{n}\left[a_i k_{s,i}f_{sd,i}A_i\right]}{\sum_{i=1}^{n}k_{s,i}f_{sd,i}A_i} \tag{3.105}$$

Example 3.14

A reinforced concrete beam carrying a design moment of 100 kNm is designed as shown below. Determine its strength fire rating. Grade of concrete is M40. Diameter of tension reinforcement is 16 mm while that of compression reinforcement is 12 mm.

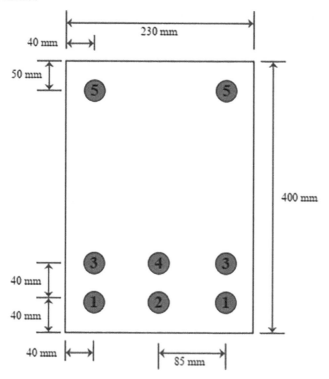

Detailed calculations are provided in the Matlab code reproduced here. The code also produces temperature contours, but those are not included in the text. If one runs the code, the contours can be seen. There is an auxiliary utility function that this code uses to compute the location of the 500°C isotherm. It is also given and will be utilized in other examples as well.

```
ex_conc_beam1.m
addpath('./mat_props/');
% Given data
b    = 230e-3;              %beam width, m
D    = 400e-3;              %beam depth, m
 fck = 40e6;                  %concrete strength, Pa
fcd = min(1 - (fck/1e6 - 50)/200, 1)*fck;    %design strength, Pa
fsd = 415e6;                %tension rebar strength, Pa
fscd = fsd;                 %compression rebar strength, Pa
 Mdemand = 120;             %kNm
 % rebar coordinates. the lower-right corner of the beam is (0,
0)
% third column is diameter of the rebar
% tension rebars
rebar_t = [ 40    40    16; ...
            125   40    16; ...
            210   40    16; ...
            40    80    16; ...
            125   80    16; ...
            210   80    16]*1e-3;
% compression rebars
rebar_c = [ 40    350   12; ...
            210   350   12]*1e-3;
 % Thermal analysis of the section
ts = 0.1:0.25:4;            %modeled time up to 4 hours
xs = 5e-3:5e-3:b/2;           %modeled width, m
ys = 5e-3:5e-3:D;             %modeled depth, m
[X, Y] = ndgrid(xs, ys);
Txy = 20*ones(length(xs), length(ys), length(ts));
Tf = 20+345*log10(480*ts+1);
 ns = 1 - 0.06*ts.^(-0.9);
for t=1:length(ts)
    nx = 0.172*log(ts(t)./(X.^2)) - 0.74;
    ny = 0.172*log(ts(t)./(Y.^2)) - 0.74;
    Txy(:, :, t) = (ns(t)*(nx+ny - 2*nx.*ny)+nx.*ny)*Tf(t);
end
figure, contour(X, Y, Txy(:, :, end)); colorbar; grid on; box on;
xlabel('x (m)'); ylabel('y (m)'); caxis([20 1000]);
title('Temperature (Celsius) at t = 1 hour')
 % structural calculations
figure, hold on
x500 = zeros(length(ts), 1);
y500 = x500;
Mcap = x500;
for t=1:length(ts)
    % find 500 isotherm
    xx500 = contour(X, Y, Txy(:, :, t), [500 500], 'linewidth', 2,
'color', 'k');
    if isempty(xx500)
        x500(t) = 0;
        y500(t) = 0;
    else
```

```
      [x500(t), y500(t)] = locate_x500(xx500(:, 2:end), b, D/2);
   end
      % fire dimensions of the beam
   bfire = b - 2*x500(t);
   Dfire = D - y500(t);
   % rebar temperatures and strength degradation factors
   % tension
   Ns_tension = size(rebar_t, 1);   %number of tension rebars
   Ts_tension = zeros(Ns_tension, 1);
   ks_tension = Ts_tension;
   As_tension = Ts_tension;
   a_tension  = Ts_tension;
   for i=1:Ns_tension
      rebar_x = rebar_t(i,1);
      rebar_y = rebar_t(i,2);
      if rebar_x>b/2
         rebar_x = b - rebar_x;
      end
      nx = 0.172*log(ts(t)./(rebar_x^2)) - 0.74;
      ny = 0.172*log(ts(t)./(rebar_y^2)) - 0.74;
      Ts_tension(i) = (ns(t)*(nx+ny - 2*nx.*ny)+nx.*ny)*Tf(t);
       ks_tension(i) = rebar_strength_reduction(Ts_tension(i));
      As_tension(i) = 0.25*pi*rebar_t(i, 3)^2;
      a_tension(i)  = rebar_t(i, 2);
   end
   % compression
   Ns_compr = size(rebar_c, 1);   %number of compression rebars
   Ts_compr = zeros(Ns_compr, 1);
   ks_compr = Ts_compr;
   As_compr = Ts_compr;
   for i=1:Ns_compr
      rebar_x = rebar_c(i,1);
      rebar_y = rebar_c(i,2);
      if rebar_x>b/2
         rebar_x = b - rebar_x;
      end
      nx = 0.172*log(ts(t)./(rebar_x^2)) - 0.74;
      ny = 0.172*log(ts(t)./(rebar_y^2)) - 0.74;
      Ts_compr(i) = (ns(t)*(nx+ny - 2*nx.*ny)+nx.*ny)*Tf(t);
      ks_compr(i) = rebar_strength_reduction(Ts_compr(i));
      As_compr(i) = 0.25*pi*rebar_c(i, 3)^2;
   end
      % strengths
   fsd_fire  = sum(ks_tension.*fsd.*As_tension)/sum(As_tension);
   fscd_fire = sum(ks_compr.*fscd.*As_compr)/sum(As_compr);
      % lever arm calculations
   a = sum(a_tension.*ks_tension.*fsd.*As_tension)/
sum(ks_tension.*fsd.*As_tension);
   dfire = D - a;
   zdash = dfire - (D - rebar_c(1, 2));
   lambda = min(0.8 - (fck/1e6 - 50)/200, 0.8);
   X_NA = sum(As_tension)*fsd_fire / (bfire * lambda * fcd);
   z = dfire - lambda*X_NA/2;
   % moment capacity calculations
   Mu1 = sum(As_tension)*fsd_fire*z;
   Mu2 = sum(As_compr)*fscd_fire*zdash;

   Mcap(t) = (Mu1+Mu2)/1e3;   %kN.m
end
```

```
 figure; hold all;
plot(ts, Mcap, 'r'); box on; grid on;
line(xlim, [Mdemand Mdemand]);
legend('Moment Capacity', 'Moment Demand')
xlabel('time (hours)'); ylabel('Moment (kNm)');
```

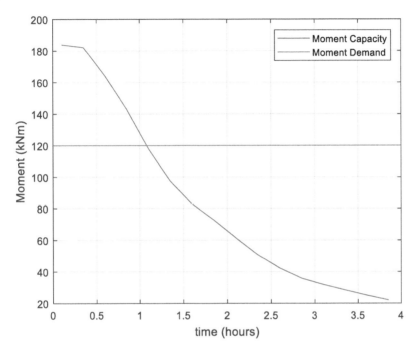

The fire rating is a little over 1 h.

locate_x500.m
```
function [x500, y500] = locate_x500(xx500, b, D)
% xx500: contour matrix having coordinates of 500 isotherm
%         (first row is x-coords, second row is y-coords)
% b    : width of beam
% D    : depth of beam
%
% x500 : location of 500 isotherm (left of vertical face)
% y500 : location of 500 isotherm (above bottom face)

tol = b*1e-3;
x500 = 1e10;
y500 = 1e10;

for n = 1:size(xx500, 2)
    if abs(xx500(1, n) - b/2)<tol
        %this will be the middle vertical edge
        y500 = min(y500, xx500(2, n));
    end
    if abs(xx500(2, n) - D)<tol
        %this will be the top edge
        x500 = min(x500, xx500(1, n));
    end
end
```

Example 3.15

A reinforced concrete beam carrying a design moment of 70 kNm is designed as shown below. Determine its strength fire rating. Grade of concrete is M40. Diameter of tension reinforcement is 16 mm while that of compression reinforcement is 12 mm.

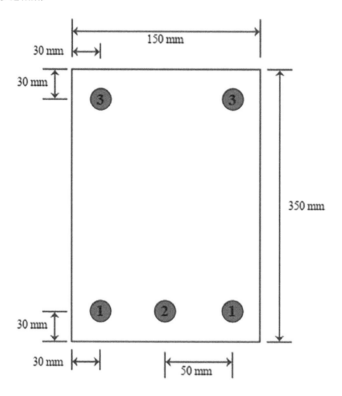

Detailed calculations are provided in the Matlab code reproduced here. The code also produces temperature contours, but those are not included in the text. If one runs the code, the contours can be seen.

ex_conc_beam2.m
```
addpath('./mat_props/');
% Given data
b    = 150e-3;              %beam width, m
D    = 350e-3;              %beam depth, m

fck = 40e6;                 %concrete strength, Pa
fcd = min(1 - (fck/1e6 - 50)/200, 1)*fck;   %design strength, Pa
fsd = 415e6;                %tension rebar strength, Pa
fscd = fsd;                 %compression rebar strength, Pa

Mdemand = 70;               %kNm

% rebar coordinates. the lower-right corner of the beam is (0, 0)
```

```
% third column is diameter of the rebar
% tension rebars
rebar_t = [ 30    30    16; ...
            80    30    16; ...
           130    30    16]*1e-3;
% compression rebars
rebar_c = [ 30   320    12; ...
           130   320    12]*1e-3;

% Thermal analysis of the section
ts = 0.1:0.25:4;              %modeled time up to 4 hours
xs = 5e-3:5e-3:b/2;           %modeled width, m
ys = 5e-3:5e-3:D;             %modeled depth, m
[X, Y] = ndgrid(xs, ys);
Txy = 20*ones(length(xs), length(ys), length(ts));
Tf = 20+345*log10(480*ts+1);

ns = 1 - 0.06*ts.^(-0.9);
for t=1:length(ts)
    nx = 0.172*log(ts(t)./(X.^2)) - 0.74;
    ny = 0.172*log(ts(t)./(Y.^2)) - 0.74;
    Txy(:, :, t) = (ns(t)*(nx+ny - 2*nx.*ny)+nx.*ny)*Tf(t);
end

figure, contour(X, Y, Txy(:, :, end)); colorbar; grid on; box on;
xlabel('x (m)'); ylabel('y (m)'); caxis([20 1000]);
title('Temperature (Celsius) at t = 1 hour')


% structural calculations
figure, hold on
x500 = zeros(length(ts), 1);
y500 = x500;
Mcap = x500;
for t=1:length(ts)
    % find 500 isotherm
    xx500 = contour(X, Y, Txy(:, :, t), [500 500], 'linewidth', 2,
'color', 'k');
    if isempty(xx500)
        x500(t) = 0;
        y500(t) = 0;
    else
        [x500(t), y500(t)] = locate_x500(xx500(:, 2:end), b, D/2);
    end

    % fire dimensions of the beam
    bfire = b - 2*x500(t);
    Dfire = D - y500(t);

    % rebar temperatures and strength degradation factors
    % tension
    Ns_tension = size(rebar_t, 1);  %number of tension rebars
    Ts_tension = zeros(Ns_tension, 1);
    ks_tension = Ts_tension;
    As_tension = Ts_tension;
    a_tension  = Ts_tension;
```

```
    for i=1:Ns_tension
        rebar_x = rebar_t(i,1);
        rebar_y = rebar_t(i,2);
        if rebar_x>b/2
            rebar_x = b - rebar_x;
        end
        nx = 0.172*log(ts(t)./(rebar_x^2)) - 0.74;
        ny = 0.172*log(ts(t)./(rebar_y^2)) - 0.74;
        Ts_tension(i) = (ns(t)*(nx+ny - 2*nx.*ny)+nx.*ny)*Tf(t);

        ks_tension(i) = rebar_strength_reduction(Ts_tension(i));
        As_tension(i) = 0.25*pi*rebar_t(i, 3)^2;
        a_tension(i)  = rebar_t(i, 2);
    end
    % compression
    Ns_compr = size(rebar_c, 1);  %number of compression rebars
    Ts_compr = zeros(Ns_compr, 1);
    ks_compr = Ts_compr;
    As_compr = Ts_compr;
    for i=1:Ns_compr
        rebar_x = rebar_c(i,1);
        rebar_y = rebar_c(i,2);
        if rebar_x>b/2
            rebar_x = b - rebar_x;
        end
        nx = 0.172*log(ts(t)./(rebar_x^2)) - 0.74;
        ny = 0.172*log(ts(t)./(rebar_y^2)) - 0.74;
        Ts_compr(i) = (ns(t)*(nx+ny - 2*nx.*ny)+nx.*ny)*Tf(t);
        ks_compr(i) = rebar_strength_reduction(Ts_compr(i));
        As_compr(i) = 0.25*pi*rebar_c(i, 3)^2;
    end

    % strengths
    fsd_fire  = sum(ks_tension.*fsd.*As_tension)/sum(As_tension);
    fscd_fire = sum(ks_compr.*fscd.*As_compr)/sum(As_compr);

    % lever arm calculations
    a = sum(a_tension.*ks_tension.*fsd.*As_tension)/
sum(ks_tension.*fsd.*As_tension);
    dfire = D - a;
    zdash = dfire - (D - rebar_c(1, 2));
    lambda = min(0.8 - (fck/1e6 - 50)/200, 0.8);
    X_NA = sum(As_tension)*fsd_fire / (bfire * lambda * fcd);
    z = dfire - lambda*X_NA/2;

    % moment capacity calculations
    Mu1 = sum(As_tension)*fsd_fire*z;
    Mu2 = sum(As_compr)*fscd_fire*zdash;

    Mcap(t) = (Mu1+Mu2)/1e3;  %kN.m
end

figure; hold all;
plot(ts, Mcap, 'r'); box on; grid on;
line(xlim, [Mdemand Mdemand]);
legend('Moment Capacity', 'Moment Demand')
xlabel('time (hours)'); ylabel('Moment (kNm)');
```

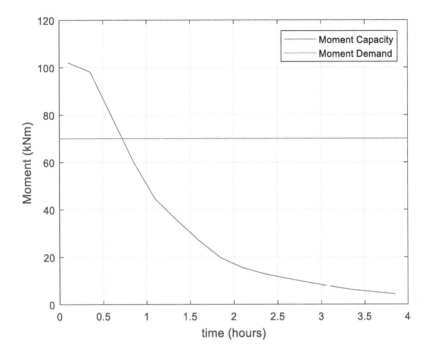

The fire rating is a about 45 min.

3.30 COMPRESSION MEMBERS – COLUMNS

3.30.1 STEEL COLUMNS

Pure compression members fail either due to crushing or buckling. The respective strengths at high temperatures can be calculated as $f_{y,fire} = k_y f_y$ and $f_{cr,fire} = k_E f_{cr}$, where f_{cr} denotes the Euler buckling stress at room temperature. The reduced load-carrying capacity can be calculated as

$$P_{fire} = \chi_{fire} A k_y f_y \qquad (3.106)$$

with appropriate material partial factor applied to f_y. χ_{fire} is the stress reduction factor depending on buckling class, slenderness ratio, and yield stress of the system given by

$$\chi_{fire} = \frac{1}{\phi + \left[\phi^2 - \lambda_{fire}^2 \right]^{0.5}}$$

$$\lambda_{fire} = \left[\frac{(k_y f_y)\left(\dfrac{L_{eff}}{r}\right)^2}{\pi^2 (k_E E)} \right]^{0.5} = \lambda \left[\frac{k_y}{k_E} \right]^{0.5}$$

$$\phi_{fire} = 0.5\left[1 + \alpha\left(\lambda_{fire} - 0.2\right) + \lambda_{fire}^2\right] \qquad (3.107)$$

Here, λ_{fire} is the temperature-dependent dimensionless slenderness ratio, L_{eff} is the effective length of the column, and α is an imperfection factor depending on buckling class of the column. Similar to the earlier discussions, the relations of reduced strength may have minor variations depending on the code. For example, the Eurocode considers a slightly different expression for ϕ_{fire} as $\phi_{fire} = 0.5\left[1 + 0.65\lambda_{fire}\sqrt{\dfrac{235}{f_y}} + \lambda_{fire}^2\right]$ with the other relations remaining the same.

When columns are subjected to bending moments in addition to axial loads, the design calculations can be suitably modified to generate temperature-dependent P-M interaction diagrams. A temperature-dependent P-M interaction diagram will shrink with the increase in temperature as a result of degradation in strength and Young's modulus of the member.

Example 3.16

A steel column having an effective length of 4.0 m is designed for a load of 1,500 kN with ISMB 450. Calculate its fire rating if it is subjected to a standard fire without any fire protection.

```
ex_steel_column1.m
addpath('./mat_props/');
% section properties
section.B  = 150e-3;       %m
section.D  = 450e-3;       %m
section.t  = 9.4e-3;       %m
section.tf = 17.4e-3;      %m
section.r  = 18.2e-2;      %m (radius of gyration)
section.Le = 4.0;          %m (effective length)


Pdemand = 1500;  %kN-m

[sf, Am, V] = section_factor(section, 'I', 4);
[sf_b, Am_b, V_b] = section_factor(section, 'boxed', 4);
ksh = 0.9*Am_b/Am;         %shielding factor

% material data:
hc = 25;         %W/(m^2-K)  (convection coefficient)
e  = 0.7;        %emissivity
fy = 250e6;      %Pa   (yield strength)
E  = 200e9;      %Pa   (Young's modulus)

% Assumed/found data
rho = 7850;      %kg/m^3    (density)
c   = 460;       %J/(kg-K)  (specific heat)
```

```
sbc = 5.67e-8; %W/(m^2-K^4) (Stefan-Boltzmann constant)
dt  = 10;       %s (chosen time increment)
tmax= 1800;     %s (time up to which temperatures are to be
calculated)

% Data structures for holding solutions
t   = 0:dt:tmax;    %time
n   = length(t);    %number of time points
T   = zeros(n, 1);  %vector to hold temperature
q   = zeros(n, 1);  %vector to hold heat flux

Ti = 25+273;  %K (initial temperature)
% Gas temperature - ISO 834
Tg = 273+25+345*log10(480*(t/3600)+1);

% Set initial values
T(1)= Ti;               %initial temperature
q(1)= hc*(Tg(1) - T(1))+ksh*e*sbc*(Tg(1)^4 - T(1)^4);

% loop to calculate successive values of temperature
for i=1:n-1
   c = steel_specific_heat(T(i)); %temperature-dependent specific
heat
   q_coeff = dt*sf/(rho*c);       % A*dt/(rho*c*V); d = A/V
   T(i+1) = T(i)+q_coeff * q(i);
   q(i+1)= hc*(Tg(i+1) - T(i+1))+ksh*e*sbc*(Tg(i+1)^4
- T(i+1)^4);
end

% calculate strength reduction factor
ky = steel_yield_strength_reduction(T);
kE = steel_modulus_reduction(T);

% calculate reduced strength with respect to time
% V is the area of cross-section
lambda_fire = sqrt((ky*fy).*(section.Le/section.r)^2./
(pi^2.*kE.*E));
phi_fire = 0.5*(1+0.65.*lambda_fire.*sqrt(235/
fy)+lambda_fire.^2);
chi_fire = 1./(phi_fire+sqrt(phi_fire.^2 - lambda_fire.^2));

Phot  = chi_fire.*V.*ky.*fy;

% plot
plot(t/60, Tg-273)
hold all; grid on; box on
plot(t/60, T-273)
xlabel('Time (min)'); ylabel('Temperature (Celsius)')
legend('Gas', 'Member', 'location', 'nw')

figure
plot(t/60, Phot/1000); grid on; box on
line(xlim, [Pdemand Pdemand])
xlabel('Time (min)'); ylabel('Load (kN)')
legend('Load capacity', 'Load demand')
```

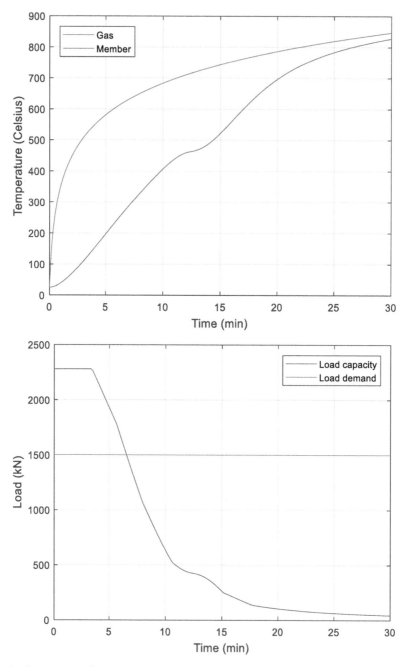

The fire rating is about 6.5 min.

Example 3.17

Calculate the fire rating of the column of Example 16 if it was protected with a 30 mm fibreboard protection.

```
ex_steel_column2.m
addpath('./mat_props/');
% section properties
section.B  = 150e-3;      %m
section.D  = 450e-3;      %m
section.t  = 9.4e-3;      %m
section.tf = 17.4e-3;     %m
section.r  = 18.2e-2;     %m (radius of gyration)
section.Le = 4.0;         %m (effective length)


Pdemand = 1500;  %kN-m

[~, Am, V] = section_factor(section, 'I', 3);
[sf_b, Am_b, V_b] = section_factor(section, 'boxed', 3);
sf = Am_b / V;           %due to fire protection
ksh = 1;                 %shielding factor

% insulation properties
din = 0.03;      %m
kin = 0.25;      %W/(m-K)

% material data:
hc = 25;         %W/(m^2-K)  (convection coefficient)
e  = 0.7;        %emissivity
fy = 250e6;      %Pa   (yield strength)
E  = 200e9;      %Pa   (Young's modulus)

% Assumed/found data
rho = 7850;      %kg/m^3    (density)
c   = 460;       %J/(kg-K)  (specific heat)
sbc = 5.67e-8;   %W/(m^2-K^4) (Stefan-Boltzmann constant)
dt  = 10;        %s (chosen time increment)
tmax= 3600;      %s (time up to which temperatures are to be
calculated)

% Data structures for holding solutions
t  = 0:dt:tmax;     %time
n  = length(t);     %number of time points
T  = zeros(n, 1);   %vector to hold temperature
q  = zeros(n, 1);   %vector to hold heat flux

Ti = 25+273;  %K (initial temperature)
% Gas temperature - ISO 834
Tg = 273+25+345*log10(480*(t/3600)+1);

% Set initial values
T(1)= Ti;              %initial temperature
q(1)= hc*(Tg(1) - T(1))+ksh*e*sbc*(Tg(1)^4 - T(1)^4);

% loop to calculate successive values of temperature
for i=1:n-1
   c = steel_specific_heat(T(i)); %temperature-dependent specific
heat
   q_coeff = (din/kin)*dt*sf/(rho*c);
   T(i+1) = T(i) +q_coeff * q(i);
   q(i+1)= hc*(Tg(i+1) - T(i+1))+ksh*e*sbc*(Tg(i+1)^4
- T(i+1)^4);
end
```

```
% calculate strength reduction factor
ky = steel_yield_strength_reduction(T);
kE = steel_modulus_reduction(T);

% calculate reduced strength with respect to time
% V is the area of cross-section
lambda_fire = sqrt((ky*fy).*(section.Le/section.r)^2./
(pi^2.*kE.*E));
phi_fire = 0.5*(1+0.65.*lambda_fire.*sqrt(235/
fy)+lambda_fire.^2);
chi_fire = 1./(phi_fire+sqrt(phi_fire.^2 - lambda_fire.^2));

Phot  = chi_fire.*V.*ky.*fy;

% plot
plot(t/60, Tg-273)
hold all; grid on; box on
plot(t/60, T-273)
xlabel('Time (min)'); ylabel('Temperature (Celsius)')
legend('Gas', 'Member', 'location', 'nw')

figure
plot(t/60, Phot/1000); grid on; box on
xlim([0 30])
line(xlim, [Pdemand Pdemand])
xlabel('Time (min)'); ylabel('Load (kN)')
legend('Load capacity', 'Load demand')
```

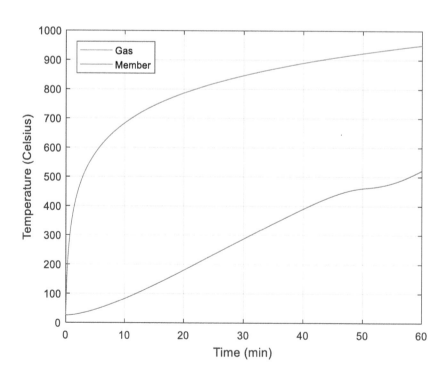

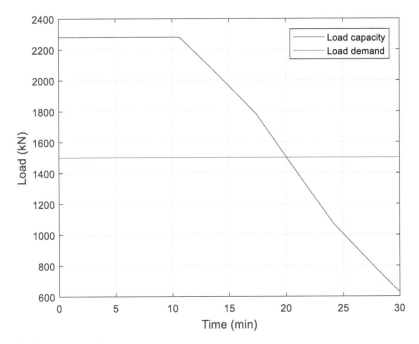

The fire rating is about 28 min.

3.30.2 REINFORCED CONCRETE COLUMNS

Design of columns under pure compression involves the consideration of reduced strength of concrete and steel reinforcement with respect to cross-sectional temperatures, similar to that of reinforced concrete beams. Employing the 500°C isotherm method, one first reduces the dimensions of the column in accordance with the 500°C isotherm, as shown in Figure 3.38.

Concrete inside the reduced dimensions is assumed to have full strength while that outside is neglected. Temperature of all the steel rebars is calculated, and their strengths are reduced accordingly. The overall compressive load capacity of the column can then be calculated as

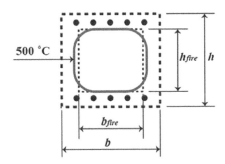

FIGURE 3.38 Consideration of b_{fire} and d_{fire} in a four-sided exposure.

$$N_{c,fire} = b_{fire} D_{fire} f_{ck}$$

$$N_{s,fire} = \frac{\sum_{i=1}^{n} \left[k_{s,i} f_{scd,i} A_i \right]}{\sum_{i=1}^{n} A_i}$$

$$N_{fire} = N_{c,fire} + N_{s,fire} \tag{3.108}$$

where $N_{c,fire}$ and $N_{s,fire}$ are the load-carrying capacities of concrete and steel, respectively. Other terms have similar meanings as introduced in the design of reinforced concrete beams.

Under combined loading conditions, the calculation approach of beams and columns can be utilized to generate temperature-dependent P-M interaction curves, and the design capacities can be calculated accordingly. Effects of buckling, if needed, can also be incorporated following a similar approach.

Example 3.18

Determine the strength rating of a short reinforced concrete column of dimensions 500 × 400 mm. Grade of concrete is M50, and Fe 500 grade rebars of 20 mm diameter are used. Load demand is 8,000 kN. Re-orient the Matlab code to perform calculations for the figure shown as well.

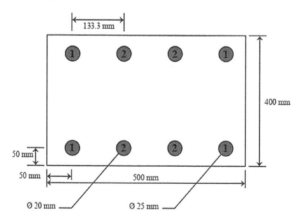

```
ex_conc_column1.m
addpath('./mat_props/');
% Given data
b    = 500e-3;              %beam width, m
D    = 400e-3;              %beam depth, m

fck = 50e6;                 %concrete strength, Pa
fsd = 500e6;                %tension rebar strength, Pa
fscd = fsd;                 %compression rebar strength, Pa

Ndemand = 8000;             %kN

% rebar coordinates. the lower-right corner of the beam is (0, 0)
% third column is diameter of the rebar
```

```
rebar = [ 50.00    50    20; ...
          183.33    50    20; ...
          316.60    50    20; ...
          450.00    50    20; ...
           50.00   350    20; ...
          183.33   350    20; ...
          316.60   350    20; ...
          450.00   350    20]*1e-3;


% Thermal analysis of the section
ts = 0.1:0.25:4;              %modeled time up to 4 hours
xs = 5e-3:5e-3:b/2;              %modeled width, m
ys = 5e-3:5e-3:D/2;              %modeled depth, m
[X, Y] = ndgrid(xs, ys);
Txy = 20*ones(length(xs), length(ys), length(ts));
Tf = 20+345*log10(480*ts+1);


ns = 1 - 0.06*ts.^(-0.9);
for t=1:length(ts)
   nx = 0.172*log(ts(t)./(X.^2)) - 0.74;
   ny = 0.172*log(ts(t)./(Y.^2)) - 0.74;
   Txy(:, :, t) = (ns(t)*(nx+ny - 2*nx.*ny)+nx.*ny)*Tf(t);
end

figure, contour(X, Y, Txy(:, :, end)); colorbar; grid on; box on;
xlabel('x (m)'); ylabel('y (m)'); caxis([20 1000]);
title('Temperature (Celsius) at t = 1 hour')


% structural calculations
figure, hold on
x500 = zeros(length(ts), 1);
y500 = x500;
Ncap = x500;
for t=1:length(ts)
   % find 500 isotherm
   xx500 = contour(X, Y, Txy(:, :, t), [500 500], 'linewidth', 2,
'color', 'k');
   if isempty(xx500)
       x500(t) = 0;
       y500(t) = 0;
   else
       [x500(t), y500(t)] = locate_x500(xx500(:, 2:end), b, D/2);
   end

   % fire dimensions of the beam
   bfire = b - 2*x500(t);
   Dfire = D - 2*y500(t);

   % rebar temperatures and strength degradation factors
   % tension
   Ns_rebar = size(rebar, 1);   %number of rebars
   Ts_rebar = zeros(Ns_rebar, 1);
   ks_rebar = Ts_rebar;
   As_rebar = Ts_rebar;
   for i=1:Ns_rebar
      rebar_x = rebar(i,1);
      rebar_y = rebar(i,2);
```

```
      if rebar_x>b/2
         rebar_x = b - rebar_x;
      end
      if rebar_y>D/2
         rebar_y = D - rebar_y;
      end
      nx = 0.172*log(ts(t)./(rebar_x^2)) - 0.74;
      ny = 0.172*log(ts(t)./(rebar_y^2)) - 0.74;
      Ts_rebar(i) = (ns(t)*(nx+ny - 2*nx.*ny)+nx.*ny)*Tf(t);

      ks_rebar(i) = rebar_strength_reduction(Ts_rebar(i));
      As_rebar(i) = 0.25*pi*rebar(i, 3)^2;
   end

   % strengths
   fscd_fire = sum(ks_rebar.*fscd.*As_rebar)/sum(As_rebar);

   % axial load capacity calculations
   Nc_fire = bfire * Dfire * fck;
   Ns_fire = sum(As_rebar) * fscd_fire;

   Ncap(t) = (Nc_fire+Ns_fire)/1e3;   %kN
end

figure; hold all;
plot(ts, Ncap, 'r'); box on; grid on;
line(xlim, [Ndemand Ndemand]);
legend('Load Capacity', 'Load Demand')
xlabel('time (hours)'); ylabel('Axial Load (kN)');
```

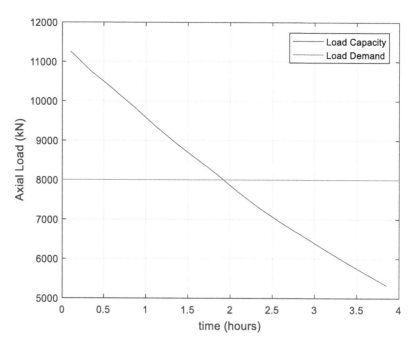

The fire rating is a little less than 2 h.

Example 3.19

Determine the strength rating of a short reinforced concrete column of dimensions 450×450 *mm. Grade of concrete is M40, and four Fe 500 grade rebars of 20 mm diameter are used with their centroids located at a distance 40 mm from the edges of the column. Load demand is 5,000 kN.*

```
ex_conc_column2.m
addpath('./mat_props/');
% Given data
b    = 450e-3;              %beam width, m
D    = 450e-3;              %beam depth, m

fck = 40e6;                 %concrete strength, Pa
fsd = 500e6;                %tension rebar strength, Pa
fscd = fsd;                 %compression rebar strength, Pa

Ndemand = 5000;             %kN

% rebar coordinates. the lower-right corner of the beam is (0, 0)
% third column is diameter of the rebar
rebar = [ 40    40    20; ...
          410   40    20; ...
          40    410   20; ...
          410   410   20]*1e-3;

% Thermal analysis of the section
ts = 0.1:0.25:4;            %modeled time up to 4 hours
xs = 5e-3:5e-3:b/2;         %modeled width, m
ys = 5e-3:5e-3:D/2;         %modeled depth, m
[X, Y] = ndgrid(xs, ys);
Txy = 20*ones(length(xs), length(ys), length(ts));
Tf = 20+345*log10(480*ts+1);

ns = 1 - 0.06*ts.^(-0.9);
for t=1:length(ts)
    nx = 0.172*log(ts(t)./(X.^2)) - 0.74;
    ny = 0.172*log(ts(t)./(Y.^2)) - 0.74;
    Txy(:, :, t) = (ns(t)*(nx+ny - 2*nx.*ny)+nx.*ny)*Tf(t);
end

figure, contour(X, Y, Txy(:, :, end)); colorbar; grid on; box on;
xlabel('x (m)'); ylabel('y (m)'); caxis([20 1000]);
title('Temperature (Celsius) at t = 1 hour')


% structural calculations
figure, hold on
x500 = zeros(length(ts), 1);
y500 = x500;
Ncap = x500;
for t=1:length(ts)
    % find 500 isotherm
    xx500 = contour(X, Y, Txy(:, :, t), [500 500], 'linewidth', 2,
'color', 'k');
    if isempty(xx500)
```

```
      x500(t) = 0;
      y500(t) = 0;
   else
      [x500(t), y500(t)] = locate_x500(xx500(:, 2:end), b, D/2);
   end

   % fire dimensions of the beam
   bfire = b - 2*x500(t);
   Dfire = D - 2*y500(t);

   % rebar temperatures and strength degradation factors
   % tension
   Ns_rebar = size(rebar, 1);   %number of rebars
   Ts_rebar = zeros(Ns_rebar, 1);
   ks_rebar = Ts_rebar;
   As_rebar = Ts_rebar;
   for i=1:Ns_rebar
      rebar_x = rebar(i,1);
      rebar_y = rebar(i,2);
      if rebar_x>b/2
         rebar_x = b - rebar_x;
      end
      if rebar_y>D/2
         rebar_y = D - rebar_y;
      end
      nx = 0.172*log(ts(t)./(rebar_x^2)) - 0.74;
      ny = 0.172*log(ts(t)./(rebar_y^2)) - 0.74;
      Ts_rebar(i) = (ns(t)*(nx+ny - 2*nx.*ny)+nx.*ny)*Tf(t);

      ks_rebar(i) = rebar_strength_reduction(Ts_rebar(i));
      As_rebar(i) = 0.25*pi*rebar(i, 3)^2;
   end

   % strengths
   fscd_fire = sum(ks_rebar.*fscd.*As_rebar)/sum(As_rebar);

   % axial load capacity calculations
   Nc_fire = bfire * Dfire * fck;
   Ns_fire = sum(As_rebar) * fscd_fire;

   Ncap(t) = (Nc_fire+Ns_fire)/1e3;   %kN
end

figure; hold all;
plot(ts, Ncap, 'r'); box on; grid on;
line(xlim, [Ndemand Ndemand]);
legend('Load Capacity', 'Load Demand')
xlabel('time (hours)'); ylabel('Axial Load (kN)');
```

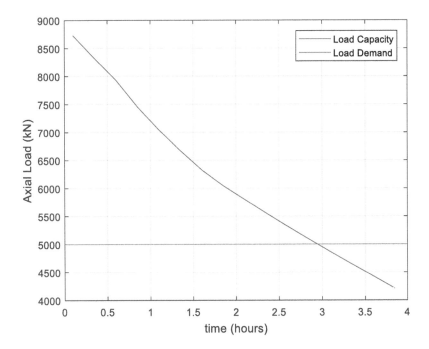

The fire rating is a little less than 3 h.

3.31 SLABS

Design of reinforced concrete slabs proceeds in a manner similar to that of beams. It is typically designed for a 1 m strip ($b = 1$m). The main difference lies in the thermal analysis. Slabs can be analysed using 1D heat transfer method (heat transfer across the thickness of the slab). This is illustrated through the following examples.

Example 3.20

A simply supported slab of span 6.5 m and thickness 200 mm is constructed with M30 grade concrete and 16 mm diameter Fe415 rebars (placed at 125 mm c/c spacing) with a clear cover of 15 mm. The bending moment demand for a 1 m strip is 70 kNm. Determine its fire rating if it is subjected to a standard fire from the bottom.

```
ex_conc_slab1.m
addpath('./mat_props/');
% Given data
b    = 1000e-3;              %slab strip of 1m width
D    = 200e-3;              %slab thickness, m

fck = 30e6;                 %concrete strength, Pa
fsd = 415e6;                %main rebar strength, Pa
sv  = 125e-3;               %main rebar spacing, m
Dbar= 16e-3;                %rebar diameter, m
cc  = 15e-3;                %clear cover, m
cbar= cc+Dbar/2;            %centroid of rebar from bottom face, m
As  = (b/sv)*pi*(Dbar/2)^2; %area of steel

Mdemand = 70;               %kNm

% Thermal analysis of the 1m strip (bottom surface is at x=0
coordinate)
ts = 0.1:0.25:4;            %modeled time up to 4 hours
xs = cbar;                  %centroid plane of rebars, m

Tx = 20*ones(length(xs), length(ts));
Tf = 20+345*log10(480*ts+1);

ns = 1 - 0.06*ts.^(-0.9);
for t=1:length(ts)
   nx = 0.172*log(ts(t)./(xs.^2)) - 0.74;
   Tx(:, t) = ns(t)*nx*Tf(t);
end

% structural calculations
for t=1:length(ts)
   % rebar temperatures and strength degradation factors
   ksy = rebar_strength_reduction(Tx(t));

   % force taken by rebars
   Fs = ksy * fsd * As;

   % force taken by concrete (divided by neutral axis depth)
   lambda = 0.8; eta = 1.0;
   Fc_by_XNA = lambda*eta*fck*b;

   % neutral axis depth
   X_NA = Fs/Fc_by_XNA;

   Mcap(t) = Fs*(D - cbar - lambda*X_NA/2)/1e3;   %kN.m
end

figure; hold all;
plot(ts, Mcap, 'r'); box on; grid on;
line(xlim, [Mdemand Mdemand]);
legend('Moment Capacity', 'Moment Demand')
xlabel('time (hours)'); ylabel('Moment (kNm)');
```

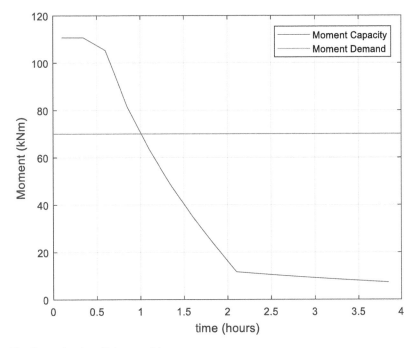

The fire rating is a little over 1 h.

Example 3.21

A simply supported slab of span 4 m and thickness 200 mm is constructed with M25 grade concrete and 18 mm diameter Fe415 rebars (placed at 125 mm c/c spacing) with a clear cover of 15 mm. The bending moment demand for a 1 m strip is 50 kNm. Determine its fire rating if it is subjected to a standard fire from the bottom.

ex_conc_slab2.m
```
addpath('./mat_props/');
% Given data
b    = 1000e-3;             %slab strip of 1m width
D    = 200e-3;              %slab thickness, m

fck = 25e6;                 %concrete strength, Pa
fsd = 415e6;                %main rebar strength, Pa
sv   = 125e-3;             %main rebar spacing, m
Dbar= 18e-3;               %rebar diameter, m
cc   = 15e-3;              %clear cover, m
cbar= cc+Dbar/2;          %centroid of rebar from bottom face, m
As   = (b/sv)*pi*(Dbar/2)^2;  %area of steel

Mdemand = 50;              %kNm

% Thermal analysis of the 1m strip (bottom surface is at x=0
coordinate)
ts = 0.1:0.25:4;           %modeled time up to 4 hours
xs = cbar;                 %centroid plane of rebars, m
```

```
Tx = 20*ones(length(xs), length(ts));
Tf = 20+345*log10(480*ts+1);

ns = 1 - 0.06*ts.^(-0.9);
for t=1:length(ts)
    nx = 0.172*log(ts(t)./(xs.^2)) - 0.74;
    Tx(:, t) = ns(t)*nx*Tf(t);
end

% structural calculations
for t=1:length(ts)
    % rebar temperatures and strength degradation factors
    ksy = rebar_strength_reduction(Tx(t));

    % force taken by rebars
    Fs = ksy * fsd * As;

    % force taken by concrete (divided by neutral axis depth)
    lambda = 0.8; eta = 1.0;
    Fc_by_XNA = lambda*eta*fck*b;

    % neutral axis depth
    X_NA = Fs/Fc_by_XNA;

    Mcap(t) = Fs*(D - cbar - lambda*X_NA/2)/1e3;   %kN.m
end

figure; hold all;
plot(ts, Mcap, 'r'); box on; grid on;
line(xlim, [Mdemand Mdemand]);
legend('Moment Capacity', 'Moment Demand')
xlabel('time (hours)'); ylabel('Moment (kNm)');
```

The fire rating is a little over 1.5 h.

4 Fire Protection

SUMMARY

Fire protection is based on a set of well-laid principles that already exist. It is vital that they should be followed without any deviations. A set of case studies are discussed in this chapter, which is important to understand the fire protection methods and practices. This chapter is coauthored by **Mr. Abhay Purandare**, who is a fire and life safety consultant practicing out of Ahmedabad. Valuable contributions to the contents and case study analyses are based on his professional experience, which is sincerely acknowledged by the authors. Summaries drawn from the case studies are views of the authors and not derived from any reports; therefore, they should be interpreted only in the academic interest and not for any legal conclusions.

4.1 FIRE PROTECTION

In the case of fire protection, it is not necessary for each generation to rediscover the principles of safety; a well-established set of principles already exist, and anyone can implement them without any deviations. Case histories help understand the safety principles and practices and evolve good practices and regulations. In developed countries, there exist concerted efforts to understand the causes of major accidents, their effects and impacts, and the need for changes in practices and regulations, if any. A few case studies are discussed in the successive sections, based on which a set of meaningful conclusions are drawn. A set of good examples can be useful to improve the fire and life safety of future occupants.

4.2 ACTIVE FIRE PROTECTION SYSTEMS

Systems and devices which require thermal, electrical, or manual triggering to control a fire are grouped under active fire protection systems. These include heat/smoke detectors, sprinklers, smoke alarms, fire extinguishers, and egress paths. Heat/smoke detectors get activated by heat, smoke, or other product of combustion and alert the occupants by light or noise signals. Smoke alarms are the most common and effective fire protection systems installed in buildings. In modern systems, mainly used in the high-rise, these detectors are designed to alert the control room of the building and the local fire station. The sprinkler systems are activated automatically by the ceiling smoke plume of fire. The primary objective of sprinkler systems is to extinguish a fire in its early stages when it is small in size. Along with sprinklers, modern high-rise buildings are also equipped with smoke management systems to exhaust the smoke from burning floors and prevent smoke accumulation along the egress paths and detectors to close the fire doors automatically.

DOI: 10.1201/9781003328711-4

An egress path (also called as exit path) is a continuous, unobstructed passage from any location within the building to an exit, leading the occupants toward the outside of the building. It includes corridors, stairways, and aisles designed according to the standard provisions leading to the exit doors. Egress or exit paths are critical for evacuating occupants and providing entry for firefighters during a fire breakout. Most of the international building codes recommend a minimum of two staircases for a high-rise building. Of course, this recommendation also depends on the occupant load and travel distances to the exit. The key idea behind active fire protection systems is their need for activation, which depends on their location relative to the fire zone, smoke movement direction, detector sensitivity, and system reliability. According to the National Fire Protection Association (NFPA), a fatal risk is reduced by half when smoke alarms are effective. It is further reduced to 80% in the presence of automatic fire sprinkler systems. In general, active systems are more effective in the early stages of a fire breakout, when it is possible to control and extinguish the fire. These systems may be rendered ineffective at later stages of fire because of higher temperatures and heat release rates (HRRs). The primary application of these systems is also helpful in maintaining the tenable conditions within the structure and prolonging the evacuation time of the occupants.

4.3   PASSIVE FIRE PROTECTION SYSTEMS

Passive fire protection systems include inherent features or systems which are derived from built-in properties of structural or non-structural members. It also includes materials that do not require any activation but are always present and ready to offer their functioning to control fire. The primary objective of the passive systems is to limit the size of fire by preventing the spread of hot gases and flames beyond the compartment. This is achieved through insulation while the integrity criteria to maintain the load-bearing capacity of the structural members at elevated temperature is assessed through stability studies. Passive systems are broadly divided into two categories: compartmentation and structural integrity. Such fire resistance systems minimize property loss. Therefore, structural members are designed to resist high temperatures with minimal strength degradation, and it depends upon the materials and structural system type used in construction. Thus, passive systems are effective in controlling the later stages of fire (fully developed fires), as structural and non-structural systems are designed to resist high temperatures.

Compartmentation aims to prevent the spread of hot gases and flames with the help of horizontal partition members (slab and floor) and vertical partition members (walls) to limit the fire spread. It is achieved in the compartment barriers by providing fire resistance in terms of insulation and integrity. Compartmentation features help to delay the fire growth to adjacent compartments or buildings. Structural integrity maintains the load-bearing capacity and stability of structural members in fire break out. It is achieved by providing fire resistance to the members in terms of their ability to resist the destructive effects of a fire. When structural members and systems are exposed to fire, they experience a rise in temperature across their cross section deteriorating their mechanical properties. It eventually causes their failure either by a partial or complete collapse of the structural systems. The constituent material of the structural member decides the type of deterioration.

4.4   CASE STUDIES

Case studies discussed in this section refer to a few accidents reported in the literature. Lessons learned from these accidents are the interpretation of the authors and do not form a part of the accident investigation report. The statements derived by the authors are meant only for academic purposes and do not testify to the findings of the investigating agencies who reported these accidents. The summary is drawn from each case study in line with the suggestions and guidelines for fire protection.

Disclaimer: The analyses given below are not formal reports of the incidents and are not meant to be used for formal/legal purposes. These analyses are based on information available in the media/research publications (which have been carefully scrutinized for relevance and authenticity), visits to the incident spot (in the case Takshashila Arcade incident), and eye-witness accounts. The collected information, along with the author's experience and knowledge, has been used to reconstruct and analyze the incidents.

4.5   STATION NIGHTCLUB, USA

A fire accident was reported in the Station Nightclub, Rhode Island, the United States of America in February 2003. It resulted in approximately a hundred fatalities, which is an alarming figure for a developed country, and hence initiated a detailed accident investigation. Prominent issues identified included egress arrangements and the fast, unmitigated growth of fire. The investigation included recreating a model for understanding fire growth and burning it down, material behavior testing, and comparing results with fire and evacuation modeling. As a result, the successive edition of NFPA 101 (Life Safety Code) was revised with respect to its egress arrangements and lining materials. It also emphasized sprinklers as mandatory for such premises.

4.6   GRENFELL TOWER, LONDON, UK

Another relevant and recent incident is the Grenfell Tower fire of 2017 in London, where external insulation and cladding got ignited after the fire flashed over in an apartment; flames began projecting outside the window openings. The combustible insulation and ACP sheets allowed the fire to spread externally covering almost the entire building and causing about 72 fatalities. The resulting study and report (which incidentally was in the public domain) recommended far-reaching changes not only for the fire safety of building materials but also for ensuring accountability and transparency in the process of fire safety design, installation, and construction. Proper understanding of the causes and effects of such incidents helps stakeholders avoid mistakes and improve relevant knowledge, resulting in safer and more resilient buildings. However, the fact is that detailed accident data are rarely available, and the practice of making accident reports public to all the concerned stakeholders is still not prevalent in many countries. Consequently, fire and life safety–related information fails to reach those people, whose jobs and lives could be made more safe and secure with this knowledge. It is also important to note that guidelines for accident reporting in case of fire do not exist in many developing countries. There is still a strong degree of hesitation to share such information due to fear of negative publicity. But it is necessary to learn correct practices of reporting fire accidents and making them available to all the stakeholders as it helps revise the design practices and fire safety guidelines.

4.7 PLASCO BUILDING, TEHRAN, IRAN

On January 19, 2017, an accidental fire, suspected to have started by an electrical fault, resulted in a major outbreak of fire. Later, it caused the total collapse of a 16-storeyed Plasco Building in Tehran, located in the capital of Iran (Yarlagadda, 2018). It is one of the important commercial centers in the city and is considered a landmark structure. Twenty-two personnel, including 16 firefighters, were killed in that incident. According to one press report, a large number of persons (about 200) were injured, and a huge financial damage (about $500 million) was projected. It was also reported that a large number of shops and businesses were not insured as the building did not comply with the safety standard requisites (Behrouz, 2019).

The Plasco building was one of the tall buildings of Tehran, consisting of a 5-storey podium and a 16-storey tower, including a basement. It is a steel structure with concrete floor slabs, categorized as a mixed-use type. It housed a shopping center, clothing workshops, and offices. While the lower floors and the podium hosted a supermarket, the upper floors mostly had offices. However, as time progressed, most of the building was reported to be occupied by garment manufacturing workshops. It greatly increased the fire load inside the building. Almost all reports are unanimous in stating that there were neither active nor passive fire protection systems in the building. Extinguishers and sprinklers were largely absent and the steel members did not have any kind of passive protection.

4.7.1 Fire Development and Growth

The fire is believed to have started on the 10th floor of the building due to an electrical short circuit, early in the morning around 8.00 am local time. As there were only a few people in the building, the fire remained unnoticed till it grew to generate a sufficient quantity of smoke. Most of the reports interpreted that the rapid spread of fire on the floor was due to stock of the material and lack of compartmentation. At least one paper (Behnam, 2019) reported severe fuel loading on floors (estimated at 1900 MJ/m²). Since sufficient oxygen is available through various openings, the fire could grow quickly and flash over in at least some compartments. Reports mentioned that flames could be seen projecting out from the floor windows. Due to lack of vertical compartmentation, hot smoke and convected heat would have traveled up through shafts and ignited the stored materials. At peak of the incident, five floors were involved, burning intensely.

By the time the Fire Service responded to the incident, the 10th floor was completely covered with smoke. At the beginning, the external firefighting could control the fire on this floor, but the fire grew and began to travel upward. Reports stated that Fire Service used aerial ladders and elevated platforms to apply external water jets to cool the material inside. Firefighting efforts continued for almost three and a half hours before the collapse of the building began.

4.7.2   Structural Collapse

Even as firefighting operations were going on, the building collapsed about three and a half hours after ignition. It was a progressive collapse; upper floors collapsed first, which was followed by the lower ones. As reported, the most probable cause of the progressive failure was the collapse of the floor beam at 12th floor level (failure of the girder-to-column connection). The impact of accumulating weight of the collapsing floor on its lower floors caused successive floors to collapse. Detailed structural analysis of the progressive failure, which was carried out later for investigation revealed an important lesson. When a floor beam is axially restrained in the presence of large deflection caused under fire, connections experience a catenary force at high temperatures. As these beam-column connections are not designed for such catenary force, failure occurred and caused a progressive collapse. It drew the attention of the structural engineers to consider such interaction forces in the analysis and probable revision to the design standards and guidelines.

4.7.3   Important Observations

This incident was very important from structural fire safety and fire protection perspective, and certain issues were very obvious.

- Structural adequacy in terms of fire resistance is critical to ensure its integrity during fire. Unprotected steel members shall fail quickly under fire and can cause progressive failure.
- Change of occupancy in a building needs to be very carefully evaluated. It can increase the fire load density manifold, causing a longer duration and more intense fire.
- Lack of compartmentation (both vertical and horizontal) allowed the fire to spread horizontally and also to the upper floors quickly.
- Absence of sprinklers and other fire protection systems increased the severity of the damage. Had sprinklers been installed, fire would have been controlled early and extinguished easily by fire services. The lack of sprinklers allowed the fire to grow unhindered.
- Housekeeping and fire safety management become critical, especially when the buildings grow older.

4.8   CHENNAI SILKS BUILDING, CHENNAI, INDIA

Similar to the Plasco building fire and collapse, a fire accident occurred in the southern metro city of Chennai. The building involved was a famous silk market called the Chennai Silks building, which housed the major brands' outlets as well as other saree showrooms and jewelry outlets. As reported, the fire which began in the morning of May 31, 2017, continued for almost 24 hrs unabated as Fire services struggled to control the incident. There was no loss of life as the building remained largely unoccupied at the time of the incident, but resulted in a huge financial loss. Parts of some floors of the building began collapsing on the following day, while the rest collapsed during the demolition of the building.

4.8.1 Building Description and Use

The Chennai Silks is a framed reinforced concrete structure with exterior and interior masonry walls, and a mixed glass facade. The building had a kitchen and a dining facility on the top floor for the employees. For fire analysis purposes, the building can be categorized as a mercantile occupancy as it housed showrooms of silks sarees and other textile items and gold jewelry. With the kind of open designs followed in such buildings, it is likely that there was no compartmentation, which would have allowed the convected heat and hot smoke to travel to all exposed floors and affected materials therein.

4.8.2 Fire Protection

It is reported that the building had a large number of portable extinguishers. As the fire was discovered at a much later stage, portable extinguishers would have been of no effective use. The presence of dry or wet risers is reported, but their effective use during fire control is not highlighted in the reports. In the absence of compartmentation, fire could have spread quickly between the floors as there was no barrier to the hot smoke and heat between floors. Openings in terms of service shafts, internal stairs, and floor openings would have aided the upward growth of the fire.

4.8.3 Fire Development and Growth

Reports stated that fire was initiated in the early morning from a generator, located in the basement due to some malfunction. Initially, it appeared that the fire had been brought under control by 10 hrs by the firefighting agencies, but got reignited and flared up again. It is important to note that silk has a heat of combustion similar to that of cotton (approximately 19–21 MJ/Kg). Due to a large quantity of material stacked vertically, which is a common practice in fabric stores, it would have resulted in a very high fire load density; this was subsequently corroborated by long duration of fire. Firefighters tried to create openings in the exterior walls to create a vent for the smoke, but these efforts were not effective. Due to the congested layout and lack of access to firefighting vehicles, externally applied water jets could not reach the seat of fire.

4.8.4 Structural Failure

During the entire duration of firefighting operations, sounds of materials or interior elements falling/collapsing were reported. Due to the intense and prolonged burning, the structure was subjected to high temperature. Most sustained building fires will experience temperature in the range of 800–900°C. Concrete begins to spall when exposed to such high temperatures. It reduces the protective layer, exposing the reinforcing steel bars to the elevated temperature. At approximately 650°C, steel has only half its yield strength. Being weak in tension, the RC floor slabs began sagging and fail. It is important to note that as the upper floors fail, they cause an additional (surcharge) loads on the lower floors, leading to a domino-like effect.

4.8.5   IMPORTANT OBSERVATIONS

Certain observations and learnings from this incident are summarized:

- This incident highlighted the importance of considering the fire load density for specifying fire resistance of structural elements. Proper fire safety management is critical in ensuring that fire loads within buildings do not exceed safe levels.
- Provision of sprinklers in buildings should be mandatory for buildings of certain occupancies considering fire load density.
- Proper compartmentation can play an important role in restricting the spread of fire, thus allowing more effective firefighting response from Fire Services.
- Provision of proper setbacks and emergency access may have allowed Fire Services to launch a more effective response and control the fire in initial stages.
- Periodic audits of electrical systems by competent persons could have helped avoiding this incident.

4.9   TAKSHASHILA COMPLEX, SURAT, INDIA

A fire occurred on the evening of May 24, 2019, around 4 pm, at a commercial complex in the Sarthana area of Surat (Purandare, 2016). The fire was controlled and extinguished by the Fire Service in the next couple of hours, but later searches confirmed 22 fatalities.

4.9.1   BUILDING DESCRIPTION AND USE

Takshashila Arcade 30×15 m is located in the Sarthana area of Surat and has G+3 floors. The building is oriented north to south lengthwise and has roads on the south and west sides. There is a single stair located on the northwest corner of the building. A road on the south side of the complex is the main Sarthana Jakat Naka-Kamrej Road, along with the Sarthana flyway (which gives a clear width of over 12 m on this side). Road on the west side is a 6 m-wide road with the Sarthana nature park and zoo on the other side of the road. There is a commercial complex on the east side (with a gap of more than 6 m) while the building abuts a two-storey villa on the north side. From a fire brigade access point of view, the south and west sides of the complex are easily accessible. The building is a commercial complex having different shops (hardware, hair cutting saloon, mobile shops) on the ground, first, and second floors; a medical clinic and nursing home/laboratory on the first floor; and training institutes on the second and third floors. The building has mixed occupancy consisting of mercantile, business, educational, and institutional occupancies. The third and fourth floors were being used by training institutes. The classification of training institutes belonging to the category B-2 (NBC, 2016), and states that their occupancy is about 100 in number.

4.9.2   EGRESS ARRANGEMENTS

There exists a single stair, which is located on the northwest corner of the building and exits directly on the west side road. The stairs are naturally ventilated, and hence it was considered that there is no possibility of accumulation of smoke. But it was observed that the stairs were smoke-logged during the incident. The staircase of 1.25 m wide extends from the ground floor to the third floor and is of concrete construction with a metal railing. An external, steel staircase is also present on the southeast corner (front side) of the building providing access only to the nursing home on the first floor. Furthermore, the electric meters for the tenants are located below the staircase on the ground floor, while the cables for the upper floors passed through a duct above the stair entrance. Building codes (NBC, Part 4) define a building above 15 m as a high-rise building. While a conventional G+3 building would not exceed 15 m in height, the addition of the fourth floor at a later stage necessitated a re-classification.

4.9.3 FIREFIGHTING ACCESS AND FIRE PROTECTION SYSTEMS

There is sufficiently wide Fire brigade access from two sides of the building to carry out external firefighting and rescue operations. However, access to the internal space is from the only stairs on the northwest side. Fire protection arrangements within the complex seem to have been limited to portable extinguishers only. Fixed fire protection systems (such as dry riser with hose reels and fire brigade inlet) are not reported to be present. In strict compliance with the building bye-laws, a high-rise mercantile building would require a wet riser, fire pumps and storage tanks, an automatic fire alarm system, and a yard hydrant. As the original status of the building does not fall under the high-rise category, such mandatory requirements were not planned originally. Hence, change of occupancy and building type causes serious violations as per the building bye-laws and also imposed severe fire risk.

4.9.4 FIRE GROWTH AND DEVELOPMENT

It is reported that the fire began around 4 pm on May24. It was reported that the fire originated from an electric fault near the staircase entrance. It is important to note that this space houses an AC compressor unit as well. The electrical duct cover, which was covered with plastic laminates also ignited and resulted in the falling of the burning material near the stair entrance. At least four two wheelers parked in front of the shop got ignited as the heat reached fuel vapors leaking from the pipe connections. Once the plastic banner adjacent to the stairs and electric cable duct on the other side of the stairs got ignited, it allowed flames to travel upward. Except for the shutters of the two shops on the ground floor, located adjacent to the stairs, other shops on the first floor and institutes on the second and third floors showed minimal evidence of damage from the fire. It appears that the material stored near the stairs on the fourth floor was ignited by the hot smoke traveling up the stairs. Flames were visible on the fourth floor, but not on the second and third floors. Construction on the terrace appears to be non-Reinforced Cement Concrete (RCC) (synthetic materials) also contributed to the growth of the fire.

The fire originating from the staircase generated intense smoke, which was pushed into the stair due to the wind direction prevalent at that time. The hot smoke traveled vertically upward through the stair shaft till it was diverted horizontally onto the fourth floor. It happened due to the vertical barrier created by the temporary roof provided on the stairs, which subsequently was burnt in the fire. This allowed the hot smoke to enter and pyrolyze the materials on the fourth floor.

4.9.5 EFFECTS ON OCCUPANTS

Smoke from combustion or pyrolysis can quickly endanger occupants in two basic ways, namely by reducing the visibility due to the suspension of solid and liquid particles in smoke and by the adverse effects of different toxic gases on the human body. Furthermore, it can also cause burns by the hot gases, but likely to occur at a later stage. But, visibility or toxicity effects can occur in the early stages of the fire. Fire gases are divided into two major groups: asphyxiants and irritants. Asphyxiants (such as carbon monoxide (CO) and hydrogen cyanide (HCN)) cause central nervous system depression, resulting in loss of consciousness and death. Carbon monoxide is always present in varying proportions in fire involved carbon fuels. Irritant gases such as halogen acids (HCl, HBr), nitrogen oxides, and ammonia enter the nose, mouth, and throat causing a burning sensation and secretion of mucus. This causes swelling of tissues, contraction of the windpipe, and leads to pulmonary edema causing death. However, it must be appreciated that much before these effects, just the presence of hot, irritant smoke is enough to cause panic among occupants. This is most likely what happened when smoke began entering the classroom.

As the fire plume began traveling horizontally toward the fourth floor (aided by the low height of the ceiling), smoke density began increasing. Soon after, as irritant effects of smoke must have kicked in, causing irritation of the eyes and the respiratory tract, occupants might have tried moving to the exit. Realizing that the smoke was coming up from the stairs, they would have rushed toward the windows on the other side in the hope of getting fresh air. As smoke quantity and temperature increased, it resulted in the loss of visibility, asphyxiation, irritation, and high heat exposure. The reaction of many occupants to jump from the fourth floor is the result of this deadly exposure. Occupants who could not reach the windows for fresh air would have been overcome by the asphyxiant effects of gases. Those who fell unconscious further inhaled these toxic gases, thus increasing the dose in their body. Charring due to heat would actually be a subsequent effect; in most likelihood, occupants were already dead by then.

4.9.6 IMPORTANT OBSERVATIONS

It is imperative that the lessons learned from this incident are not forgotten and are used to improve fire safety for our future generations:

- It was proved once again that smoke (visibility and toxicity effects) is the leading cause of deaths in buildings. Importantly, even a small size fire can generate a considerable amount of smoke, which can jeopardize the safety of occupants.
- It must be understood that egress design is of paramount importance in buildings. Hence, all international codes emphasize equally on egress design as fire protection of buildings. Even if firefighting and rescue operation

is delayed, occupants must be able to safely escape from the building in the event of emergency. For commercial buildings, minimum of two exits should be made mandatory, irrespective of height.

- Older buildings, which are non-compliant from egress design point of view, need to be checked for the design of stairs and probable fire hazards.
- Modifications and changes to building design need to be carefully evaluated from a fire risk perspective. Concerned (Municipal/Fire) officials need to be trained and certified to carry out fire and life safety assessments for building designs.
- Implementation of building rules and code provisions should be in spirit and not just for compliance.
- Fire and life safety features should be maintained throughout the working life of the building.
- Strengthening and upgrading fire services is very vital. Current shortages in Fire Service infrastructure, staff, and training are too large to expect them to deliver services effectively. Role of experts and consultants need to be considered for fire and life safety design and installation and inspections in buildings as fire services are currently not equipped for this role.
- Sensitization of society toward the issue of fire and life safety is an important step. Fire safety has to be a people movement; simple enforcement is not a solution.
- Students must be taught about fire safety in schools so that younger and future generation is conscious about fire and life safety issues.

References

accessengineeringlibrary.com. 2022. Website accessed on June 2022.

AIChE, American Institute of Chemical Engineers. 1996. *Guidelines for evaluating process plants for external explosions and fire*, Center for Chemical Process Safety, New York.

Arturson, G. 1987. The tragedy of San Juanico—the most severe LPG disaster in history, *Burns Including Thermal Injury*, 13, 87–102.

ASCE. 2005. *Standard calculation methods for structural fire protection*, Standard ASCE 29, New York.

Assael, M.J., Kakosimos, K.E. 2010. *Fire, explosion and toxic gas dispersions: Effects, calculation and risk analysis*, CRC Press, Boca Raton, FL, New York.

ASTM. 2020. *Standard test methods for fire tests of building construction and materials. E119*, American Society for Testing and Materials, New York.

ASTM. 2022. *Standard test methods for determining effects of large hydrocarbon pool fires on structural members and assemblies. E1529*, American Society for Testing and Materials, New York.

Aven, T., Vinnem, J.E. 2007. *Risk management with applications to offshore petroleum industry*, Springer, London, ISBN: 9781846286520, pp. 212.

Baalisampang, T., Abbassi, R., Garaniya, V., Khan, F., Dadashzadeh, M. 2018. Review and analysis of fire and explosion accidents in maritime transportation. *Ocean Engineering*, 158, 350–366.

Babrauskas, V. 1981. A closed form approximation for post flashover compartment fires. *Fire Safety Journal*, 4, 67–73.

Babrauskas, V., Grayson, S.J. (Eds.). 1992. *Heat release in fires*. Elsevier Applied Science, Amsterdam.

Behnam, B. 2019. Fire structural response of the plasco building: A preliminary investigation report. *International Journal of Civil Engineering*, 17, 563–580.

Bergman, T.L., Lavine, A.S., DeWitt, D.P., Incropera, F.P. 2011. *Introduction to heat and mass transfer*, 6th edition, John Wiley & Sons, USA.

Bermingham, M.J., Kent, D., Zhan, H., St John, D.H., Dargusch, M.S. 2015. Controlling the microstructure and properties of wire arc additive manufactured Ti–6Al–4V with trace boron additions. *Acta Materialia*, 91, 289–303. 10.1016/j.actamat.2015.03.035

BIS. 1979. *IS 3809: Fire resistance tests for structures*, Bureau of Indian Standards, India.

Brode, H.L. 1959. Blast wave from a spherical charge. *The Physics of Fluids*, 2 (2), 217–229.

BS EN 10025-2. 2019. *Hot rolled products of structural steels. Technical delivery conditions for non-alloy structural steels*, BSI, UK.

BS EN 10088-4. 2009. *Stainless steels: Technical delivery conditions for sheet/plate and strip of corrosion resisting steels for construction purposes*, BSI, London

BS EN 1993-1-1. 2005. *UK National Annex to Eurocode 3: Design of steel structures General rules and rules for buildings*, BSI, UK.

Buchanan, A.H. (Ed.). 2001. *Fire engineering design guide*, 2nd edition, Centre for Advanced Engineering, University of Canterbury, New Zealand.

Buchanan, A.H., Abu, A.K. 2017. *Structural design for fire safety*. Wiley, Hoboken, NJ.

Budnick, E.K., Evans, D.D., Nelson, H.E. 1997. *Simple fire growth calculations, Chapter 10, NFPA Fire Protection Handbook*, 18th edition, National Fire Protection Agency, Quincy, Massachusetts.

Casal, J., Salla, J.M. 2006. Using liquid superheating energy for a quick estimation of overpressure in BLEVEs and similar explosions. *Journal of Hazardous Materials*, 137 (3), 1321–1327.

CCPS. 1989. *Center for Chemical Process Safety: Guidelines for chemical process quantitative risk analysis*, AIChE, New York.

CCPS. 1994. *Center for Chemical Process Safety: Guidelines for evaluation characteristics of vapour could explosion, flash fire and BLEVE*, AIChE, New York.

CCPS. 2010. *Guidelines for vapor cloud explosion, pressure vessel burst, BLEVE, and flash fire hazards*, 2nd edition, Wiley Subscription Services, Inc., A. Wiley Company, New York.

CEN. 2002. *Eurocode 1: Basis of design and design actions on structures, Part 2-2. EN 1991-2-2, Brussels*, European Committee for Standardization, Brussels.

CEN. 2004. *Eurocode 2: Design of concrete structures, Part 1–2: General rules – structural fire design. EN 1992-1-2, Brussels*, European Committee for Standardization, Brussels.

CEN. 2005. *Eurocode 3: Design of steel structures, Part 1–2: General rules – structural fire design. EN 1993-1-2, Brussels*, European Committee for Standardization, Brussels.

Chamberlain, G.A. 1987. Developments in design methods for predicting thermal radiation from flares. *Chemical Engineering Research Design*, 65, 299–309.

Chandrasekaran, S. 2010. Risk assessment of offshore pipelines. Keynote address at *HSE in Oil and gas- exploration and production, International HSE Meet*, IBC-Asia, Kuala Lumpur, Malaysia, December 6–8.

Chandrasekaran, S. 2014. *Advanced toffshore plant FEED engineering*, Changwon National University Press, Republic of South Korea, pp. 237. ISBN: 978-89-969-7928-9.

Chandrasekaran, S. 2015. *Advanced marine structures*, CRC Press, Florida (USA), ISBN: 978-14-987-3968-9.

Chandrasekaran, S. 2016. *Offshore structural engineering: Reliability and risk assessment*, CRC Press, Florida, ISBN: 978-14-987-6519-0.

Chandrasekaran, S. 2016. *Health, safety and environmental management for offshore and petroleum engineers*, John Wiley and Sons, UK. ISBN: 978-11-192-2184-5.

Chandrasekaran, S. 2017. *Dynamic analysis and design of ocean structures*, 2nd edition, Springer, Singapore. ISBN: 978-981-10-6088-5.

Chandrasekaran, S. 2019. *Advanced steel design of structures*. CRC Press, Florida. ISBN: 978-036-72-3290-0.

Chandrasekaran, S. 2020a. *Design of marine risers with functionally graded materials*, Woodhead Publishing, Elsevier, pp. 200, ISBN: 9780128235379.

Chandrasekaran, S. 2020b. *Offshore semi-submersible platform engineering*, CRC Press, Florida, pp. 240, ISBN: 9780367673307.

Chandrasekaran, S., Hari, S. 2022. Functionally graded materials and their application to marine structures. *International Journal of Sustainable Marine Structures*, 04 (01), 35–41.

Chandrasekaran, S., Hari, S., Amirthalingam, M. 2019. Wire-arc additive manufacturing of functionally-graded material for marine riser applications, Proceeding of International Conference on OCEAN 2019, Universiti Malaysia, Terengganu, Malaysia, 5–7th August 2019.

Chandrasekaran, S., Hari, S., Murugaiyan, A. 2020. Wire arc additive manufacturing of functionally graded material for marine risers. *Journal of Materials Science, & Engineering*, A, 792–139530.

Chandrasekaran, S., Hari, S., Murugaiyan, A. 2022. Functionally graded materials for marine risers by additive manufacturing for high-temperature applications: Experimental investigations. *Structures*, 35, 931–938.

Chandrasekaran, S., Jain, A.K. 2016. *Ocean structures: Construction, materials, and operations*, CRC Press, Florida, ISBN: 978-149-87-9742-9.

Chandrasekaran, S., Jain, A.K., Shafiq, N., Mubarak, M., Wahab, A. 2021. *Design aids for offshore platforms under special loads*, CRC Press, Florida, pp. 280, ISBN: 9781032136844.

Chandrasekaran, S., Kiran, A. 2014b. Consequence analysis and risk assessment of oil and gas industries. *Proceedings of International Conference on Safety & Reliability of Ship, Offshore and subsea structures*, Glasgow, UK, August 18–20.

Chandrasekaran, S., Kiran, A. 2014a. Accident modelling and risk assessment of oil and gas industries. *Proceedings of 9th Structural Engineering Convention*, pp. 2533–2543, IIT, Delhi, India. December 22–24.

Chandrasekaran, S., Kiran, A. 2015. Quantified risk assessment of LPG Filling station, Professional safety. *Journal of American Society of Safety Engineers*, 44–51.

Chandrasekaran, S., Kumar Bhattacharyya, Subrata. 2012. *Analysis and Design of Offshore Structures with illustrated examples. Human Resource Development Center for Offshore and Plant Engineering (HOPE Center)*, Changwon National University Press, Republic of Korea, ISBN: 978-89-963-9155-5, pp. 285.

Chandrasekaran, S., Nagavinothini, R. 2020. Behaviour of stiffened deck plates under hydrocarbon fire. *Journal of Marine Systems & Ocean Technology*, 15, 95–109, DOI: 10.1007/s40868-020-00077-1.

Chandrasekaran, S., Pachaiappan, S. 2020. Numerical analysis and preliminary design of topside of an offshore platform using FGM and X52 Steel under special loads, *Journal of Innovative Infrastructure Solutions*, DOI: 10.1007/s41062-020-00337-4.

Chandrasekaran, S., Srivastava, G. 2017. *Design aids for offshore structures under special environmental loads including fire resistance*, Singapore ISBN 978-981-10-7608-4.

Chang, Y.F., Chen, Y.H., Sheu, M.S., Yao, G.C. 2006. Residual stress-strain relationship for concrete after exposure to high temperatures, *Cement and Concrete Research*, 36 (10), 1999–2005, DOI: 10.1016/j.cemconres.2006.05.029.

Chen, X., Li, J., Cheng, X., He, B., Wang, H., Huang, Z. 2017. Microstructure and mechanical properties of the austenitic stainless steel 316L fabricated by gas metal arc additive manufacturing. *Materials Science and Engineering: A*, 703, 567–577. 10.1016/j.msea.2017.05.024

cityfire.co.uk (website, accessed June 2022).

Crowl, D.A. 1991. Using thermodynamic availability to determine the energy of explosion. *Plant/Operations Progress*, 10(3), 136–142.

Crowl, D.A. 1992a. Using thermodynamic availability to determine the energy of explosion for compressed gases. *Plant/Operations Progress*, 11 (2), 47–49.

Crowl, D.A. 1992b. Calculating the Energy of explosion using thermodynamic availability, *Journal of Loss Prevention and Safety*, 109–118.

Crowl, D.A., Louvar, J.F. 2002. *Chemical process safety: Fundamentals with applications*, Prentice Hall, New Jersey, ISBN: 0-13-0181765.

ctif.org (website, accessed May 2022).

Dadashzadeh, M., Khan, F., Hawboldt, K., Amyotte, P. 2013. An integrated approach for fire and explosion consequence modelling. *Fire Safety Journal*, 61, 324–337.

Delichatsios, M.A. 1984. Flame height of turbulent wall fire with significant flame radiation. *Combustion Science and Technology*, 39, 195–214.

Drysdale, D. 1998. *An introduction to fire dynamics*, 2nd edition, John Wiley & Sons, Chichester, UK.

Dubnikova, F., Kosloff, R., Almog, J., Zeiri, Y., Boese, R., Itzhaky, H., Alt, A., Keinan, E. 2005. Decomposition of Triacetone Triperoxide Is an Entropic Explosion. *Journal the American Chemical Society*, 127 (4), 1146–1159.

Elia, R. 1991. *Risk assessment and management for chemical process industry*, H.R. Greenberg and J.J. Cramer (Eds), Van Nostrand Reinhold, New York.

english.www.gov.cn (website, accessed May 2022).

Feasey, R., Buchanan, A.H. 2002. Post-flashover fires for structural design. *Fire Safety Journal*, 37 (1), 83–105.

Gao, J.W., Wang, C.Y. 2000. Modeling the solidification of functionally graded materials by centrifugal casting. *Materials Science and Engineering: A*, 292 (2), 207–215. 10.1016/S0921–5093(00)01014–5.

Hall, J.R. 2014. *The total cost of fire in the United States*, National Fire Protection Association, Quincy, MA.

Hasemi, Y., Tokunaga, T. 1983. Modeling of turbulent diffusion flame and fire plumes for analysis of fire growth, *Proceeding of 21st National Heat Transfer Conference*, American Society of Mechanical Engineers, Seattle, Washington, July 24–28.

Hasemi, Y., Tokunaga, T. 1984. Some experimental aspects of turbulent diffusion flames and buoyancy plumes for fire sources against a wall and in corners of walls. *Combustion Science and Technology*, 40, 1–17.

Hemmatian, B., Planas, E., Casal, J. 2015. Fire as a primary event of accident domino sequences: the case of BLEVE. *Reliability Engineering and System Safety*, 139, 141–148.

Hemmatian, B., Planas, E., Casal, J. 2017. Comparative analysis of BLEVE mechanical energy and overpressure modelling. *Process Safety and Environmental Protection*, 106, 138–149.

Hurley, M.J., Gottuk, D.T., Hall Jr., J.R., Harada, K., Kuligowski, E.D., Puchovsky, M., Torero, J.L., Watts Jr., J.M., Weiczorek, C.J. (Eds.). 2015. *SFPE Handbook of fire protection engineering*, Springer Nature, United Kingdom, ISBN: 9781493925643.

IBC. 2015. *International Building Code*. International Codes Council, NJ, USA.

Idogaki, T. 1979. On the Specific Heat of Fe(HCOO)2·2H2O. *Journal of the Physical Society of Japan*, 47 (2), 498–504, DOI: 10.1143/JPSJ.47.498.

ISO. 2012. Fire resistance tests – Elements of building construction. ISO 834–2012. International Organization for Standardization.

Jin, G., Takeuchi, M., Honda, S., Nishikawa, T., Awaji, H. 2005. Properties of multi-layered mullite/Mo functionally graded materials fabricated by powder metallurgy processing. *Materials Chemistry and Physics*, 89 (2–3), 238–243. 10.1016/j.matchemphys.2004.03.031

Kawagoe, K. 1958. *Fire behavior in rooms. Report No. 27*, Building Research Institute, Tokyo, Japan.

Khan, F.I., Abbasi, S.A. 1999. Major accidents in process industries and analysis of causes and consequences. *Journal of Loss Prevention in Process Industries*, 12, 361–378.

Khan, N., Srivastava, G. 2018. Enhanced fire severity in modern Indian dwellings. *Current Science*, 115 (2), 320–325.

Kodur, V. 2014. Properties of concrete at elevated temperatures, *ISRN Civil Engineering*, DOI: 10.1155/2014/468510.

Law, M. 1983. A basis for the design of fire protection of building structures. *The Structural Engineer*, 61A, 1.

Lees, F.P. 1996. *Loss prevention in chemical process industries*, Butterworth-Heinemann, London.

Lie, T.T. 1995. *Fire temperature-time relations. SFPE Handbook of Fire Protection Engineering*, 2nd edition, Society of Fire Protection Engineers, USA.

Lin, Y-S. 2005. Estimations of the probability of fire occurrences in buildings. *Fire Safety Journal*, 40 (8), 728–735.

Magnusson, S.E., Thelandersson, S. 1970. Temperature-time curves of complete process of fire development. *Bulletin of Division of Structural Mechanics and Concrete Construction*, 16(16), 1–181.

Martina, F., Mehnen, J., Williams, S.W., Colegrove, P., Wang, F. 2012. Investigation of the benefits of plasma deposition for the additive layer manufacture of Ti–6Al–4V. *Journal of Materials Processing Technology*, 212 (6), 13.

MBIE. 2007. *New Zealand building code handbook*, 3rd edition, Wellington, New Zealand.

Naser, M.Z., Kodur, V.K.R. 2015. A probabilistic assessment for classification of bridges against fire hazard. *Fire Safety Journal*, 76, 65–73.

National Building Code. 2016. *Part 4 – Fire & Life Safety*, Bureau of Indian Standards, New Delhi, India

NFPA. 1997. *Fire protection handbook*, 18th edition, National fire Protection Association, Quincy, MA.

NFPA Report. 2020. *Fire loss in the US*. National Fire Protection Association, Quincy, MA.

Ogle, R.A., Ramirez, J.C., Smyth, S.A. 2011. Calculating the explosion energy of a boiling liquid expanding vapor explosion using exergy analysis. *Process Safety Progress*, 31 (1), 51–54.

Pan, J., Zou, R., Jin, F. 2017. Experimental study on specific heat of concrete at high temperatures and its influence on thermal energy storage. *Energies*, 10 (1), DOI: 10.3390/en10010033.

Pape, R. 1988. Calculation of intensity of thermal radiation from large fire. *Loss Prevention Bulletin*, 82, 1.

pinkerton.com (website, accessed May 2022).

Planas Cuchi, E., Salla, J.M., Casal, J. 2004. Calculating overpressure from BLEVE explosions. *Journal of Loss Prevention Process Industries*, 17 (6), 431–436.

Planas, E., Casal, J. 2015. BLEVE-Fireball. *Handbook of Combustion*, 1 (21), 1–25. Wiley VCM, Weinheim.

Prugh, R.W. 1991a. Quantitative evaluation of "BLEVE" hazards. *Journal of Fire Protection Engineering*, 3 (1), 9–24.

Prugh, R.W. 1991b. Quantify BLEVE hazards, *Chemical Engineering Progress*, 87, 66.

Purandare, A. 2016. The escalating fire hazards of modern homes. *The Fire Engineer*, January–March, 2016.

Purser, D. 2002. *Society of Fire Protection Engineers (SPFE) Guide to Human Behavior*, 3rd edition, Springer, Maryland.

Quintiere, J.G. 2016. *Principles of fire behaviour*, 2nd edition, Taylor and Francis, ISBN: 9781498735629.

Rajan, T.P.D., Pillai, R.M., Pai, B.C. 2010. Characterization of centrifugal cast functionally graded aluminum-silicon carbide metal matrix composites. *Materials Characterization*, 61 (10), 923–928. 10.1016/j.matchar.2010.06.002.

Razus, D., Molnarne, M., Movileanu, C., Irimia, A. 2006. Estimation of Limiting oxygen concentration of fuel-air-inert mixtures at elevated temperatures by means of adiabatic flame temperature. *Journal of Chemical Engineering Processing: Process Intensification*, 45 (3), 193–197.

Reid, R.C. 1979. Possible mechanism for pressurized-liquid tank explosions or BLEVE, *Science*, 3, 203.

reuters.com website. 2021. https://www.reuters.com/article/us-iran-building-idUSKBN1550JE

Roberts, A.F. 1982. Thermal radiation hazards from release of LNG fire from pressurized storage. *Fire Safety Journal*, 4, 197.

Roberts, H.C.W. 1952. Report on the causes of, and circumstances attending, the explosion occurred at Essington Colliery, County Durham, Her Majesty's Stationery Office, London, 9, 39–40.

Shishkovsky, I., Missemer, F., Smurov, I. 2012. Direct metal deposition of functionally graded structures in the Ti-Al system. *Physics Procedia*, 39, 382–391. 10.1016/j.phpro.2012.10.052

Smith, J.M., Van Ness, H.C., Abbott, M.. 1996. *Introduction to chemical engineering thermodynamics*, 5th edition. McGraw-Hill, Inc., New York.

Srivastava, G., Gandhi, P. 2022. *Performance of combustible façade systems used in green building technologies under fire*, Springer-Nature, Singapore.

Tellez, C., Pena, J.A. 2002. Boiling-liquid expanding-vapour explosion: An introduction to consequence and vulnerability analysis. *Safety and Chemical Engineering Program Newsletter, American Institute of Chemical Education*, 206–211.

The, O. 1934. Thermal conductivity of irons and steels. *Journal of the Franklin Institute*, 217 (5), 619–620. DOI: 10.1016/s0016-0032(34)90359-8.

Thomas, P.H., Heselden, A.J.M. 1972. Fully developed fires in single compartments, CIB Report No 20. Fire Research Note 923, Fire Research Station, UK.

Tong, Shu-jiao, Lo, Siu-ming, Zhong, Pei-hong, Chen, Bao-zhi. 2013. Jet fire consequence evaluation on the natural gas transported by pipelines. *Procedia Engineering*, 52, 349–354.

Übeyli, M., Balci, E., Sarikan, B., Öztaş, M.K., Camuşcu, N., Yildirim, R.O., Keleş, Ö. 2014. The ballistic performance of SiC–AA7075 functionally graded composite produced by powder metallurgy. *Materials & Design (1980–2015)*, 56, 31–36. 10.1016/j.matdes.2013.10.092

Udayavani.com, 2019. Website: *https://www.udayavani.com/english-news/surat-fire-tragedy-lapses-on-part-of-civic-body-builder-finds-preliminary-probe,* accessed 29th May, 2019

UL. 2001. *UL 10B: Fire test of door assemblies*, Underwriters Laboratories, USA.

UL. 2012. *UL 263: Fire tests of building construction and materials*, Underwriters Laboratories, USA.

Wickstrom, U. 1985. Application of the standard fire curve for expressing natural fires for design purposes. In: Harmathy, T.Z. (Ed.), *Fire safety: science and engineering, ASTM STP 882* (pp. 145–150), American Society of Testing and Materials, PA.

Wickstrom, U. 1986. A very simple method for estimating temperature in fire exposed concrete structures. In: Grayson, S.J., Smith, D.A. (Eds.), *Proceedings of new technology to reduce fire losses and costs*. Elsevier, New York.

Wickstrom, U. 2016. *Temperature calculation in fire safety engineering*. Springer-Nature, Cham.

Wielgosz, E., Kargul, T., Falkus, J. 2014. Comparison of experimental and numerically calculated thermal properties of steels. *Proceeding 23rd International Metallurgical and Materials Conference May 2014*, pp. 1528–1533, 2014.

Wray, Harry, A. (Ed). 1992. *Manual on flashpoint standards and their use: Methods and regulations*, ASTM International, Baltimore, ISBN: 0-8031-1410-9.

Yarlagadda, T., Hajiloo, H., Jiang, L., Green, M., Usmani, A. 2018. Preliminary modelling of plasco tower collapse. *International Journal of High-Rise Buildings*, Council on Tall Building and Urban Habitat, Korea, 7(4), 397–408.

Yaws, C.L., Braker, W. 2001. *Metheson gas data book*, 7th edition, McGraw Hill Professional, ISBN 9780071358545, pp. 982.

Yeo, J.G., Jung, Y.G., Choi, S.C. 1998 Design and microstructure of ZrO2/SUS316 functionally graded materials by tape casting. *Materials Letters*, 37 (6), 304–311. 10.1016/S0167-577X(98)00111-6.

Yuan, H., Li, J., Shen, Q., Zhang, L. 2012. In situ synthesis and sintering of ZrB2 porous ceramics by the spark plasma sintering–reactive synthesis (SPS–RS) method. *International Journal of Refractory Metals and Hard Materials*, 34, 3–7. 10.1016/j.ijrmhm.2012.01.007.

Index

Note: **Bold** page numbers refer to tables and *italic* page numbers refer to figures.